ELEMENTARY
CLASSICAL
MECHANICS

ELEMENTARY CLASSICAL MECHANICS

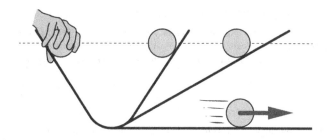

Stephen Wiggins
University of Bristol, UK

W\$ World Scientific

NEW JERSEY · LONDON · SINGAPORE · BEIJING · SHANGHAI · HONG KONG · TAIPEI · CHENNAI · TOKYO

Published by

World Scientific Publishing Co. Pte. Ltd.

5 Toh Tuck Link, Singapore 596224

USA office: 27 Warren Street, Suite 401-402, Hackensack, NJ 07601

UK office: 57 Shelton Street, Covent Garden, London WC2H 9HE

Library of Congress Control Number: 2023026154

British Library Cataloguing-in-Publication Data
A catalogue record for this book is available from the British Library.

ELEMENTARY CLASSICAL MECHANICS

ISBN 978-981-127-745-0 (hardcover)
ISBN 978-981-127-924-9 (paperback)
ISBN 978-981-127-746-7 (ebook for institutions)
ISBN 978-981-127-747-4 (ebook for individuals)

For any available supplementary material, please visit
https://www.worldscientific.com/worldscibooks/10.1142/13443#t=suppl

Typeset by Stallion Press
Email: enquiries@stallionpress.com

Printed in Singapore

Preface

This book consists of material for a one-term course in classical mechanics presented to first year mathematics majors at the University of Bristol. Unlike the undergraduate mathematics curriculum in the United States, the United Kingdom has a rich tradition of teaching classical mechanics (among other courses that might be viewed as "theoretical physics") at the undergraduate level. (This should not be too surprising since, after all, Isaac Newton did invent calculus, in part, to describe mechanical phenomena.) I have always found this a very enjoyable course to teach, but a bit challenging. The class tends to be large, it had been a requirement for all first year mathematics students, and the students have varied backgrounds. Some have had a basic course in mechanics, some have not. Almost all have a good background in calculus, but I generally need to assume minimal background in differential equations and linear algebra. Moreover, this needs to be a mathematics course, not a physics or engineering type mechanics course. So, for example, free body diagrams are kept to a bare minimum. The result is this book. The goal is to give students some exposure to the beautiful subject of classical mechanics at a level that will "mesh" with their other first mathematics courses and give them the prerequisites (and hopefully tempt them to take) more advanced courses in classical, quantum, and statistical mechanics.

While this book originated during my time in the city of Bristol, United Kingdom, I have completed it in the past few months during my stay at the United States Naval Academy in Annapolis, Maryland. I am grateful for the support of the William R. Davis '68 Chair in the Department of Mathematics at the United States Naval Academy.

Contents

Preface   v

List of Figures   ix

List of Tables   xi

Prelude: What is Mechanics and What Framework Will
  We Use to Understand it?   xiii

Chapter 1.   Kinematics — Scalars, Vectors, and Vector Algebra   1

Chapter 2.   Kinematics — Vectors and Coordinates   13

Chapter 3.   Kinematics — Space Curves, their Description
             and Derivatives, Circular Motion, and Line Integrals   27

Chapter 4.   Examples of the Computation of Line Integrals
             and Newton's Axioms   45

Chapter 5.   Dynamics — Motion of a Particle in One Dimension   55

Chapter 6.   Dynamics — Projectiles, Constrained Motion, Friction   67

Chapter 7.   Dynamics — Work and Power   81

Chapter 8.   Dynamics — Conservation of Energy and Momentum   91

Chapter 9.   Dynamics — The Phase Plane for One-Dimensional
             Motion   101

Chapter 10. Dynamics — The Simple Pendulum. Torque
and Angular Momentum 109

Chapter 11. Motion in a Central Force Field 119

Bibliography 129

Index 131

List of Figures

1.1 The definition of a vector as a directed line segment from the point P to the point Q. 2

1.2 (a) Two vectors $\mathbf{A}$ and $\mathbf{B}$. (b) The sum of $\mathbf{A}$ and $\mathbf{B}$. (c) Illustration of the parallelogram law for vector addition. 5

1.3 (a) Five vectors, $\mathbf{A}$, $\mathbf{B}$, $\mathbf{C}$, $\mathbf{D}$, and $\mathbf{E}$. (b) The sum of the five vectors, $\mathbf{F} = \mathbf{A} + \mathbf{B} + \mathbf{C} + \mathbf{D} + \mathbf{E}$. 6

1.4 The angle between two vectors. 7

1.5 Projection of $\mathbf{A}$ onto $\mathbf{B}$. 8

1.6 Illustration of the right hand rule. 9

1.7 Vectors for Problem 1. 10

2.1 The $\mathbf{i}$, $\mathbf{j}$, and $\mathbf{k}$ unit vectors in the direction of the x, y, and z axes, respectively. 14

2.2 The projection of $\mathbf{A}$ onto $\mathbf{i}$, $\mathbf{j}$, and $\mathbf{k}$. 15

2.3 Geometry of a space curve. 24

3.1 Geometry of the coordinate system associated with a space curve. 28

3.2 Motion of a particle constrained to a circle. 32

3.3 The relationship between polar coordinates and Cartesian coordinates. 33

3.4 Geometry of the path for computing the line integral. 36

4.1 The points p_1 and p_2. The dashed lines represent coordinate axes with negative values. 46

4.2 The path from p_1 to the origin along the x-axis, then from the origin to p_2 along the y-axis. 46

4.3   The path along the straight line in the $x - z$ plane from p_1 to $(0, 0, 1)$, then along the straight line in the $y - z$ plane from $(0, 0, 1)$ to p_2.   47

4.4   The path from p_1 to p_2 defined by $\frac{3}{4}$ of the unit circle in the $x - y$ plane. The perspective in this figure is awful, just think of the points in the $x - y$ plane that the unit circle must pass through.   48

4.5   Two distinct coordinate systems in three-dimensional space. . . .   52

6.1   The force of gravity acting on a particle.   68

6.2   Geometry of the motion of a projectile under the influence of gravity.   69

6.3   Geometry of the motion of a particle under the influence of gravity down an inclined plane.   72

6.4   Particle moving on a sphere under the influence of gravity. . . .   73

7.1   Geometry associated with a force acting on a particle in three-dimensional space.   82

7.2   An example of a force acting on a particle.   83

7.3   A force moving a particle around a circle.   84

7.4   Geometry of a force acting on a particle in three-dimensional space.   86

8.1   Geometry associated with a force moving a particle between two points in space.   92

8.2   A particle moving vertically under the influence of gravity. . . .   95

8.3   A particle moving on a sphere.   97

9.1   Geometry of motion associated with a level set of the energy.   103

9.2   Graphical procedure for drawing the phase portrait from the potential energy.   104

9.3   A more complicated example of the graphical procedure for drawing the phase portrait from the potential energy.   106

10.1   Geometry associated with the pendulum.   110

10.2   The potential energy and the phase portrait associated with the pendulum.   112

10.3   Geometry associated with a force acting on a particle moving in three-dimensional space.   113

11.1   Geometrical illustration of a central force.   120

11.2   Polar coordinate system associated with a particle moving in the xy plane.   121

11.3   Illustration of the Law of Areas.   122

List of Tables

4.1   Units and dimensions. 51

5.1   Examples of mathematical forms of forces for Newton's
equations in one dimension.   56

Prelude: What is Mechanics and What Framework Will We Use to Understand it?

What is Mechanics and What Framework Will We Use to Understand it?

We begin by first setting the stage for what these lectures are about. What is "mechanics"? Very simply, it is the study of *motion*. Motion of what? Well, just about anything. There are "mechanical theories" or "descriptions" for such diverse *systems* (another word to define, at some point) as the internet, DNA molecules, and the stockmarket, just to name three. So the study of mechanics will prove useful in just about every field of science, engineering, and technology, and should be viewed as essential.

But let's go back to the beginning. With our definition of mechanics we have raised the obvious question, "what is motion?" Answering this question will tell us a great deal about *how* we are going to study "mechanics".

Suppose I throw a ball across the room. In this case it should be clear what I mean by "the motion of the ball"; I am merely referring to the path the ball takes as it moves through the air on its way across the room. "Describing the motion" may mean providing a description of the ball's location at each instant of time as it moves across the room. But what do we mean by "time" and "location"? This may seem "obvious", but what is "obvious" to you, may not be so to someone else, and one of our goals will be to understand just what we can take as "obvious", and what does require more careful consideration.

Time. We will make no attempts to define "time" in this course. It is a notion that we can all reach some level of agreement based upon our common *experience*. Most importantly, we can provide a way of measuring it with clocks that we can all agree upon. "Time" is an example of a quantity referred to as a *dimension*, and dimensions have different *units* (e.g. seconds, minutes, hours). There are some deep issues lurking behind the scenes here, issues that led Einstein to develop the theory of special relativity, but you will learn about them in other courses in your studies.

Space. The ball is moving through "space". What is space? Like time, for us space will be an undefined notion. However, again like time, it will achieve meaning and utility for us by providing a way to measure it, and this will involve the notion of *length* or *distance*. So how do we describe the ball's "location in space" at a fixed instant of time? This can be done by specifying a *coordinate system*, and by specifying the location of the ball *with respect to a chosen coordinate system*. This last phrase is significant, because there is a great deal of freedom in choosing a coordinate system. "Convenience" is often the most significant motivating factor for the choice of a coordinate system. However, it should be clear that some aspects of the motion of the ball should be *independent* of the choice of coordinates used to quantitatively describe its motion. Understanding what aspects of motion are *coordinate dependent* versus *coordinate free* is important.

"Length" is another example of a *dimension*, and you should be familiar with many *units* of length (e.g. meters, inches, or miles).

The location of the ball at each instant in time is referred to as the *kinematics* of the ball. The subject of mechanics is usually broken into two parts. Kinematics, which is the study of the *description* of motion, and *dynamics*, which is the study of the *cause* of motion. Dynamics would be concerned with what caused the particular kinematics of the ball. That is, why did it take a certain path with a certain speed (among other things that we don't have the language to discuss yet).

The beginning point for the study of dynamics is "Newton's Laws". What is a "law"? It's something we cannot prove mathematically. Rather, it is the distillation of years of experience (or experiment) that we take as "the truth" until something better comes along (as a result of more experience and experiment). Newton's laws were laid down by him in his book *The Principia*[1] (the title is actually longer, but you can look that

---

[1]Isaac Newton. *The Principia: Mathematical Principles of Natural Philosophy.* Univ of California Press, 1999.

up, it is usually just referred to as "The Principia"). We next describe Newton's laws following the description of Sommerfeld.[2]

Here's the first law. (It happens that Newton's first law was arrived at many years earlier by Galileo (see the references below), but that's just the way science works sometimes.)

> **Newton's First Law.** *Every material body remains in its state of rest or of uniform rectilinear motion unless compelled by forces acting on it to change its state.*

We would like to make this law a bit more mathematical. In the *Principia* Newton included a series of definitions in order to clarify this.

> **Definition II.** *The quantity of motion is the measure of the same, arising from the velocity and the quantity of matter conjunctly.*

The notion of "velocity" should be intuitively evident from our earlier discussion, once we've chosen a coordinate system, that is. "Quantity of matter" is a different issue.

Mass. Physical objects are made up of "stuff", or matter. For us, this will be another primitive concept that we will make no effort to define. "Mass" is another example of a *dimension*, and you are probably familiar with many *units* of mass (e.g. grams and kilograms). Newton did make an effort to define mass in his Definition 1, quoted below.

> **Definition I.** *The quantity of matter is the measure of the same, arising from its density and bulk conjunctly.*
>
> Thus air of double density, in a double space, is quadruple in quantity; in a triple space, sextuple in quantity. The same thing is to be understood of snow, and fine dust or powders, that are condensed by compression or liquefaction; and of all bodies that are by any caused whatever differently condensed. I have no regard in this place to a medium, if any such there is, that freely pervades the interstices between the parts of bodies. It is this quantity that I mean hereafter everywhere under the name of body or mass. And the same is known by the weight of each body; for it is proportional to the weight, as I have found by experiments on pendulums, very accurately made, which shall be shewn hereafter.

---

[2] Arnold Sommerfeld, MO Stern, and RB Lindsay. *Mechanics.* Lectures on Theoretical Physics, Volume i, 1954.

This is a good example of "circular reasoning" (Sommerfeld refers to it as a "mock definition"), see if you can figure out why. It should give you hope to see geniuses making the same types of mistakes that you may make (but it's no excuse!).

So for Newton, "quantity of matter" and "mass" are synonyms.

According to Newton's First Law, the "quantity of motion" is the product of the mass and velocity. Now we quantify the notion of mass with just a number (e.g. 6 kilograms, 600 kilograms, etc.), but "velocity" is not just quantified by a number, but also by a *direction* (think about it, it makes a difference which way you are going). Such "directed quantities" are referred to as *vectors*, and they will be the main subject of the lectures for the next few weeks.

Now let's refer to the mass of the material body by m and the velocity by $\mathbf{v}$, we make the velocity boldface to denote that it is a directed quantity (in our context, we want to distinguish it from m). We shall refer to the quantity of motion of the material body as its *momentum*, and denote it by $\mathbf{p}$ (the phrase "quantity of motion" is not very good as it does not reflect the directional quality of the motion. The idea of a vector was not fully formed in Newton's time, though he doubtless would have understood the concept). Then we can write the equation:

$$\mathbf{p} = m\mathbf{v},$$

and Newton's first law becomes:

$$\mathbf{p} = \text{constant in the absence of forces},$$

(yes, we have not said what a force is). Sometimes this is referred to as *the Law of Inertia*.

Referring to the motivating example from above, the ball started from rest in my hand, then it moved across the room. Its "quantity of motion" changed. Therefore a force, or forces, must have "acted on it" (of course, I threw it with my arm, but there were other forces too, right? What were they?). We can't *predict* the motion without knowing the forces. *Knowing the forces acting on a body is the key issue in solving any mechanics problem.*

But let us return to this phrase "material body". In our "thought experiment" of tossing a ball across the room and describing its motion would it have made a difference if the ball was "small, round and hard" or "a large plastic bag filled with water" or "four balls connected by springs"? Probably. The point here is the structure and composition of the material

body will probably have an effect on the path it takes as it is thrown across
the room. This is a level of complexity that we will not consider at this
point. Our material bodies will consist of all of the mass being concen-
trated in a point, sometimes referred to as a "point mass" or "mass point".
Surprisingly, this *approximation* works well in many situations. For example, we can get a pretty good idea of the dynamics of the solar system by
considering the sun, planets, and some of the moons, as point masses (but
for *detailed* calculations of trajectories of objects in the solar system this
approximation would not be adequate).

Now let us turn to the "real law of dynamics"; Newton's second law (if
the second law is the "real law", then why isn't it the first law? There is a
"logical" answer. Try to figure it out.).

> **Newton's Second Law.** *The change in motion is proportional
> to the force acting and takes place in the direction of the straight
> line along which the force acts.*

Now we cannot avoid saying something about this term "force" any
longer. Intuitively, you should have some idea of this notion just from using
your muscles in daily life. From this, it should also seem reasonable that
force is another "directed quantity" or vector. We can measure these forces
that we generate with our muscles, or other machines, by comparing them
against the "standard" force of gravity. Once we specified a way of measuring a quantity, it achieves physical meaningfulness.

Now let's write out Newton's second law as an equation:

$$\frac{d}{dt}\mathbf{p} = \mathbf{F}.$$

The "change of motion" is written as the time derivative of the momentum, and this is equal to the force. What about the phrase "... direction
of the straight line along which the force acts"? How is that manifested
in this equation? You will learn that when we focus on the properties of
vectors during the next few lectures.

For completeness, we write down Newton's third law.

> **Newton's Third Law.** *Action always equals reaction, or, the
> forces two bodies exert on each other are always equal and opposite
> in direction.*

This says that forces always occur in pairs in Nature. You attract the
earth just as much as the earth attracts you. The third law will play a

central role when we couple (or, "attach together") many point masses to form more complicated bodies.

"Physical measurement" is the notion that gives meaning to many of the quantities in Newton's three laws, hence the laws themselves. But if you look back at our discussion of the different physical measurements that we described in order to quantify *time*, *space*, and *mass* underlying them all was the notion of a chosen coordinate system. Are Newton's law dependent on the coordinate system? We will learn later that, for a certain class of coordinate systems, the answer is "no".

This is the end of the general introduction. Now we will get down to developing the "nuts and bolts" of mechanics and, for the most part, forget about much of the issues that were raised here. We will merely take Newton's Laws as *axioms*, and once we've determined the forces, derive the mathematical consequences. Newton's laws will be taken up in more detail after we've developed some tools to study kinematics.

References

We mentioned that there are *many* textbooks on mechanics that cover the basic material dealt with in this course. At some point you may want to purchase a particular text for reference purposes. The particular text that you purchase may depend upon your tastes, interests, and needs that you develop after a few years study of mathematics. However, we mention below a few books that go somewhat beyond the course and whose study will yield deep insights.

The book by Truesdell[3] is a scholarly and entertainingly written work on the history of mechanics. In it you will learn of some of the intellectual struggles of some of the greatest scientific minds as they developed some of the concepts that we now almost take for granted. This is an excellent book to learn about how mathematics is applied to describe and predict physical phenomena.

Mach[4] was one of the deepest thinkers about what mechanics really "means". It is said that Einstein studied this book throughly in the time leading up to his development of relativity. It is easy to dip in and out of the book depending on your interests, and thinking about what Mach

[3]Clifford Truesdell. *Essays in the History of Mechanics*. Springer Science & Business Media, 2012.

[4]Ernst Mach. *The Science of Mechanics: A Critical and Historical Account of its Development*. Open Court Publishing Company, 1907.

writes will certainly give much insight into mechanics. This book is now available for free download at:

http://www.archive.org/details/scienceofmechani005860mbp

The book of Sommerfeld[5] is a personal favorite of mine. Sommerfeld was one of the greatest theoretical physicists of the 20th century, and also one of the greatest teachers. It is probably fair to say that this book is a bit "obscure". However, the topics, problems, and insights are simply unequalled in any "modern" textbook, and I still use it regularly as a reference.

Finally, I want to mention the *Schaum's Outline on Theoretical Mechanics* by Spiegel.[6] This is an excellent text from which to learn classical mechanics. It develops the theory at an adequate level and it contains a wealth of problems that provide additional insight. I have used it in my course and many of the problems have made their way into my material. It is very unfortunate that it has been out of print in recent years.

---

[5]Arnold Sommerfeld, MO Stern, and RB Lindsay. *Mechanics*. Lectures on Theoretical Physics, Volume i, 1954.

[6]Murray R Spiegel. *Theory and Problems of Theoretical Mechanics*, Schaum's Outline Series. *and Ex*, 3:73–74, 1967.

Chapter 1

Kinematics — Scalars, Vectors, and Vector Algebra

Broadly speaking, mechanical systems will be described by a combination of *scalar* and *vector* quantities. A scalar is just a (real) number. For example, mass or weight is characterized by a (real and nonnegative) number. A vector is characterized by a non-negative real number (referred to as a *magnitude*), *and* a direction. For example, momentum and force are examples of vector quantities, since it is not only the magnitude of these quantities that is important, but they also have a "directional nature" (for the moment, we are just appealing to your intuition here). We assume that you are familiar with the algebraic rules associated with manipulating numbers. Similar algebraic rules exist for manipulating vectors, and that is what we want to develop now (for example, it is important to know how to deal with several different forces acting on a "body" in different directions). However, we are getting a bit ahead of ourselves. First, we need to give a mathematical definition of a vector. The rules of vector algebra are then derived from the properties ascribed to vectors through the definition (the axiomatic way of reasoning).

As you progress in your studies of mathematics you will see that the notion of a vector is fundamental in many areas of studies. This is possible because the fundamental idea has been generalized by virtue of its being abstracted and axiomatized. However, the abstraction and axiomatization sprang from a fundamental, and fairly simple, idea (it's always insightful to look for the motivation behind the abstraction), and this is how we will develop the definition of a vector.

The Definition of a Vector

We now define the notion of a vector that is sufficient for the purposes of this book. In other courses you will generalize the notion of a vector, but the fundamental properties will remain the same. It is important to realize that our definition of a vector does not require reference to a specific coordinate system. It is "coordinate free". This is important to realize because later we will "represent" vectors in different coordinate systems.

Our idea of the three-dimensional space in which we live is taken as a basic, undefined concept. We consider two points in space, labelled P and Q, and the line segment connecting them, starting at P and ending at Q see Fig. 1.1.

Hence, a vector is a *directed line segment*, in the sense that its direction is from P to Q. This is denoted in the figure by the arrow on the line segment. The magnitude of the vector is just the *length* of the vector.

Our definition of a vector is valid in ordinary (three-dimensional) space. Do not let the fact that the paper on which Fig. 1.1 is drawn fool you. Work out in your mind that the definition is perfectly valid in three dimensions. Geometrical reasoning is very powerful in mathematics. Many complicated ideas can be conveyed (and, allegedly, understood) with a simple sketch. However, if the limitations of the sketch are not well understood from the beginning, then false ideas can also be conveyed. There is a bit of art work involved here; something that is gained with experience and careful thought. Some mathematicians may give you the impression that mathematics is a very "cut-and-dried" subject; that it flows logically through an orderly current of successive theorems, proofs, lemmas, propositions, and definitions. This is nonsense! Many important areas of mathematics were originally formulated in a very confused, illogical, and messy state, and it is only through many years of thought and refinement that they are brought

Fig. 1.1. The definition of a vector as a directed line segment from the point P to the point Q.

into the perfectly crisp and logical state that they are presented in today (which is also often dry and uninspiring).

Now we need to introduce some *notation* for vectors. That is, a way of representing them on a piece of paper, or a black or whiteboard. This is important. Notation forms the language of mathematics, so you must get this right from the start. In the history of mathematics many battles have been fought over the best notation for a subject (a notable such battle was the one between Newton and Leibniz (and their respective followers) for the notation of the calculus that you are learning this year).

The vector shown in Fig. 1.1 can be denoted in several ways:

$$\overrightarrow{PQ}, \quad \vec{A}, \quad \underline{A}, \quad \mathbf{A}.$$

The first notation, $\overrightarrow{PQ}$, probably makes a lot of sense from the point of view of the definition of a vector that we gave (and good mathematical notation should be intuitive in the sense that it conveys the aspects of the quantity that is being represented). Unfortunately, we will rarely use such notation. Now you are probably wondering where the "A" came from. It is certainly not shown in Fig. 1.1. It is at this point that we perform a bit of abstraction with respect to our original definition of a vector. A vector is defined by a "length and direction". So the beginning point (P in the figure) and the end point (Q in the figure) are not crucial for defining a vector; any two points will do provided that the segment between those two points has the same length and direction. Or, to put it another way, we can move the line segment all over space and, so long as we maintain its length and direction, it always defines the same vector (this will be important shortly when we talk about addition of two vectors). This makes sense because, thinking ahead, we want to use vectors to represent quantities such as momentum and force (yes, things we have not defined yet). The force acting on a body has a characteristic that is independent of the particular location in space that it is being applied. For example, you can apply the same force to lifting an object whether that object is in Bristol or in London. But back to notation. Since the two particular endpoints are not essential we economize the notation and use a single character, in this case "A" to denote a vector. On the printed page it is denoted in boldface, $\mathbf{A}$. This doesn't work so well on a black or whiteboard. In that case either an underline is used, $\underline{A}$, or an "overarrow", $\vec{A}$.[1]

---

[1] A vector has two characteristics: length (or "magnitude") and direction.

Vector Algebra

CONSISTENT WITH OUR DEFINITION OF A VECTOR, WE NOW DEFINE HOW VECTORS ARE ADDED, MULTIPLIED BY SCALARS ("NUMBERS"), AND SUBTRACTED. With the definition of a vector in hand we can now develop rules for adding vectors, subtracting vectors, and multiplying vectors by scalars (the idea of multiplying two vectors will be considered a bit later). But first, we need to establish two important properties.

Equality of Two Vectors. Given two vectors, say $\mathbf{A}$ and $\mathbf{B}$, what does it mean for these two vectors to be *equal*? Since a vector is defined by its direction *and* length, we say that $\mathbf{A}$ and $\mathbf{B}$ are equal if they have the same direction and length, and we write this as $\mathbf{A} = \mathbf{B}$. Note that equality of two vectors does not depend at all on their beginning and end points in space.

The Negative of a Vector. Consider a vector $\mathbf{A}$. The vector having the same length as $\mathbf{A}$, but with *opposite* direction, is denoted by $-\mathbf{A}$.

The Sum of Two Vectors. Now let us move on to the subject of adding two vectors. Suppose we are given two vectors $\mathbf{A}$ and $\mathbf{B}$. What meaning could we give to the sum of $\mathbf{A}$ and $\mathbf{B}$? We first place the beginning point of $\mathbf{B}$ at the ending point of $\mathbf{A}$, as shown in Fig. 1.2(b). Then the *sum of* $\mathbf{A}$ *and* $\mathbf{B}$, denoted $\mathbf{A} + \mathbf{B}$, is defined to be the vector, which we will call $\mathbf{C}$, from the beginning point of $\mathbf{A}$ to the ending point of $\mathbf{B}$, see Fig. 1.2(b). Sometimes the word *resultant* is used for *sum*. This definition is equivalent to the *parallelogram law* for vector addition, which we illustrate in Fig. 1.2(c).

It should be clear that this definition for vector addition can be easily extended to more that two vectors. We illustrate this in Fig. 1.3.

The Difference of Two Vectors. The difference of two vectors $\mathbf{A}$ and $\mathbf{B}$, denoted $\mathbf{A} - \mathbf{B}$, is the vector $\mathbf{C}$ which when added to $\mathbf{B}$ gives $\mathbf{A}$. This definition, along with the definition of the negative of a vector given above, implies that we can write $\mathbf{A} - \mathbf{B}$ as $\mathbf{A} + (-\mathbf{B})$ (it would be very instructive if you drew a "picture" of this definition using the definitions given already). If $\mathbf{A} = \mathbf{B}$ then $\mathbf{A} - \mathbf{B}$ is defined as the *zero vector*, and denoted by 0. The length of the zero vector is zero, but its direction is undefined.

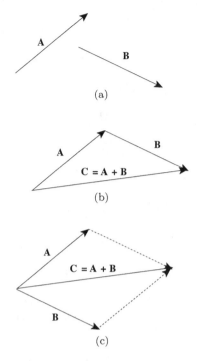

Fig. 1.2. (a) Two vectors **A** and **B**. (b) The sum of **A** and **B**. (c) Illustration of the parallelogram law for vector addition.

The Product of a Vector and a Scalar. The product of a vector **A** with a scalar c is the vector $c\mathbf{A}$ (or $\mathbf{A}c$) with magnitude $|c|$ times the magnitude of **A** and the direction the same as **A** if c is positive, or the direction opposite to **A** if c is negative. If $c = 0$ then $c\mathbf{A} = 0$, the zero vector.

Unit Vector. This is an important notion that will arise many times. For this reason we choose to highlight the idea. A *unit vector* is simply a vector having length one ("unit length"). Given any vector **A**, with length $A > 0$, we can construct a unit vector from it, as follows:

$$\frac{\mathbf{A}}{A} \equiv \mathbf{a},$$

then we can also write

$$\mathbf{A} = A\mathbf{a}.$$

For a given vector **A** we will often denote its magnitude by:

$$A = |\mathbf{A}|.$$

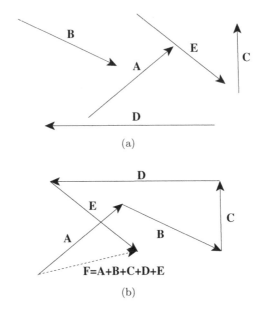

(a)

(b)

Fig. 1.3. (a) Five vectors, **A**, **B**, **C**, **D**, and **E**. (b) The sum of the five vectors, **F** = **A** + **B** + **C** + **D** + **E**.

Note that the only thing distinguishing two unit vectors is their respective directions.[2]

Kinematics–Multiplication of Vectors

THERE ARE TWO WAYS TO MULTIPLY TWO VECTORS: THE DOT (OR SCALAR) PRODUCT AND THE CROSS (OR VECTOR PRODUCT).

Now we come to the subject of *muliplication of vectors*. If you think of our purely geometrical definition of a vector given previously, it may seem rather strange to think of multiplying two "arrows in space". However, our notion of multiplication will also be based on geometry.

There are two ways in which we will multiply vectors. They arise naturally in different settings in mechanics. They are the *dot, or scalar product* and the *cross, or vector product*. This should already seem a bit novel. After all, there is only one way to multiply numbers, and when you multiply two numbers, you get a number. The terminology here seems to indicate

---

[2]Vectors are added using the "parallelogram law". This rule for addition of vectors does not require coordinates.

that, besides there being two ways to multiply vectors, if you do it one way you get a number and if you do it the other way you get a vector. This is indeed the case, but let's get to the definitions.

Dot, or Scalar Product. Let **A** and **B** be vectors. Then the *dot, or scalar product* of **A** and **B**, denoted **A** · **B**, is defined by:

> **Key point:**
>
> $$\mathbf{A} \cdot \mathbf{B} \equiv |\mathbf{A}|\,|\mathbf{B}|\cos\theta, \quad 0 \le \theta \le \pi,$$

where θ is the angle between **A** and **B** when we give **A** and **B** the same starting points, see Fig. 1.4.

We say that two vectors are *perpendicular*, if their dot product is zero. An equivalent term is "orthogonal". Two vectors are said to be *orthogonal* if their dot product is zero. A related term is "orthonormal", but this tends to be applied *exclusively* to unit vectors. Two unit vectors are said to be *orthonormal* if their dot product is zero.[3]

With the dot product of two vectors defined we can discuss the notion of "the projection of one vector onto another".

The Projection of a Vector onto Another Vector. Consider the vectors **A** and **B** as shown in Fig. 1.5. What do we mean by "the projection of **A** onto **B**?"

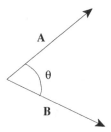

Fig. 1.4. The angle between two vectors.

---

[3]We noted earlier that we will assume certain basic knowledge and take some ideas as evident, e.g. the notions of space, time, and mass. Concerning space, the notion that two points determine a straight line, is also something that most of you should accept based on your experience with Euclidean geometry. In the hierarchy of Euclidean geometrical objects, the next most complicated object (beyond a line) would be a plane. You should understand completely, and make sure you are comfortable with, the statements three points determine a plane and two vectors determine a plane.

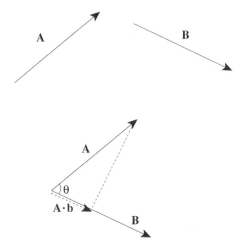

Fig. 1.5. Projection of **A** onto **B**.

Let **b** = $\frac{\mathbf{B}}{|\mathbf{B}|}$ denote the unit vector in the direction of **B**. Then the *projection of* **A** *on* **B** is given by:

$$\mathbf{A} \cdot b = |\mathbf{A}| \cos\theta.$$

Why didn't we define the projection of **A** onto **B** as just **A** · **B**?

Cross, or Vector Product. Let **A** and **B** be vectors. Then the *cross, or vector product* of **A** and **B**, denoted **A** × **B**, is defined by:

Key point:

$$\mathbf{A} \times \mathbf{B} \equiv |\mathbf{A}|\,|\mathbf{B}| \sin\theta\,\mathbf{n}, \quad 0 \le \theta \le \pi,$$

---

where θ is the angle between **A** and **B** and **n** is a unit vector perpendicular to the plane formed by **A** and **B**.

However, **n** is not uniquely defined in this way. There are two possibilities, and we will have to make a choice of one so that we are all speaking the same language.

Here's how to understand the possibilities. Take your *right hand*, and point the thumb in the direction of **A** with the fingers in the direction of **B**, when you do this consider Fig. 1.6. Then **n** could be chosen to either point "upward" out of your hand [Fig. 1.6(a)], or "downward", through your hand [Fig. 1.6(b)]. We make the first choice, i.e. that shown in Fig. 1.6(a).

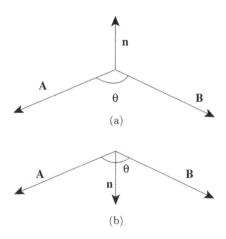

Fig. 1.6. Illustration of the right hand rule.

If $\mathbf{A} = \mathbf{B}$, or $\mathbf{A}$ and $\mathbf{B}$ are collinear (parallel), then we define $\mathbf{A} \times \mathbf{B} = 0$, the zero vector.

A Final Remark: Dimensions. We have been taking the notions of two- and three-dimensional space as primitive concepts since most people's experience makes them comfortable with these notions. However, many applications of mechanics involve the consideration of higher dimensional space. How do we define vectors in higher dimensions? The graphical constructions we have utilized here do not appear to generalize easily. This is where *abstraction* comes in, and you will learn about it in detail later in your studies.[4]

Problem Set 1

1. For the vectors shown in Fig. 1.7 construct:

(a) $-\mathbf{A} + \mathbf{B} + 2\mathbf{C}$,
(b) $\mathbf{A} - \mathbf{B} - 2\mathbf{C}$,
(c) $2\mathbf{A} + \mathbf{B} - \mathbf{C}$,
(d) $\mathbf{A} + \frac{1}{2}\mathbf{B} - \frac{1}{2}\mathbf{C}$.

---

[4]
- The dot product of two vectors is a scalar.
- The cross product of two vectors is a vector.
- The definitions of dot product and cross product do not require coordinates.

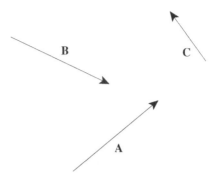

Fig. 1.7. Vectors for Problem 1.

What is the relation between the vectors you constructed in (a) and (b)? What is the relation between the vectors you constructed in (c) and (d)?

2. Suppose **A**, **B**, and **C** are vectors and a and b are scalars. Using only the definitions given in class concerning vectors, and the rules you already know for manipulating scalars, graphically demonstrate the following laws of vector algebra.

(a) $\mathbf{A} + \mathbf{B} = \mathbf{B} + \mathbf{A}$,   Commutative Law for Vector Addition,

(b) $\mathbf{A} + (\mathbf{B} + \mathbf{C}) = (\mathbf{A} + \mathbf{B}) + \mathbf{C}$,   Associative Law for Vector Addition,

(c) $a(b\mathbf{A}) = (ab)\mathbf{A} = b(a\mathbf{A})$,   Associative Law for Scalar Multiplication,

(d) $(a + b)\mathbf{A} = a\mathbf{A} + b\mathbf{A}$,   Distributive Law,

(e) $a(\mathbf{A} + \mathbf{B}) = a\mathbf{A} + a\mathbf{B}$,   Distributive Law.

3. Suppose **A**, **B**, and **C** are vectors and a is a scalar. Using the definition of the dot, or scalar product, given earlier, show that it has the following properties.

(a) $\mathbf{A} \cdot \mathbf{B} = \mathbf{B} \cdot \mathbf{A}$,   Commutative Law for Dot Products,

(b) $\mathbf{A} \cdot (\mathbf{B} + \mathbf{C}) = (\mathbf{A} \cdot \mathbf{B}) + (\mathbf{A} \cdot \mathbf{C})$,   Distributive Law,

(c) $a(\mathbf{A} \cdot \mathbf{B}) = (a\mathbf{A}) \cdot \mathbf{B} = \mathbf{A} \cdot (a\mathbf{B}) = (\mathbf{A} \cdot \mathbf{B})a$.

4. Let **A** and **B** be nonzero vectors. Prove that $\mathbf{A} + \mathbf{B}$ and $\mathbf{A} - \mathbf{B}$ are perperpendicular if and only if $|\mathbf{A}| = |\mathbf{B}|$.

Draw a diagram to show that, geometrically, this means that if **A** and **B** are the sides of a parallelogram, then the parallelogram is, in fact, a rhombus.[5]

5. Prove that the area of a parallelogram with sides **A** and **B** is $|\mathbf{A} \times \mathbf{B}|$. Argue that from this result it is easy to conclude that the area of the *triangle*, with two sides given by **A** and **B**, is $\frac{1}{2}|\mathbf{A} \times \mathbf{B}|$.[6]

---

[5]This is an important problem for understanding the geometric meaning of the dot product. Note that it does *not* require the use of any coordinate system.

[6]Similar to the previous problem, this is an important problem for understanding the geometric meaning of the cross product. It also does *not* require the use of any coordinate system.

Chapter 2

Kinematics — Vectors and Coordinates

We have emphasized that our development of the definition and properties of vectors did not utilize at all the notion of a *coordinate system*, i.e. an agreed upon set of locations with respect to which the location of a vector in space is described. In this sense a vector is a *coordinate free* concept. This is important to understand because many physical quantities should be independent of the particular system of coordinates in which they may be considered.

Nevertheless, you may have felt a certain "vagueness" with the geometrical arguments given to prove the various laws of vector algebra. Coordinates are very useful for calculating certain quantities and making certain notions more "concrete". Many of the laws of vector algebra, as well as the properties of the dot and cross products, are easy to verify using coordinates (*but*, these laws and properties are *independent* of the particular choice of coordinates). In this chapter we will consider different types of coordinate systems and how we *represent* a vector in a given coordinate system. We begin with a coordinate system with which you should already have some familiarity.

Rectangular, or Cartesian, Coordinates in Three-Dimensions. Consider three unit vectors in three-dimensional space, **i**, **j**, and **k** emanating from the same point (which we call 0, for origin) having the property that each unit vector is perpendicular to the other two, i.e.

> **Key point:**
>
> $$\mathbf{i} \cdot \mathbf{j} = 0, \, \mathbf{j} \cdot \mathbf{k} = 0, \, \mathbf{i} \cdot \mathbf{k} = 0. \qquad (2.1)$$

These three vectors define three *coordinate axes*, which are referred to as the x, y, and z axes, respectively. There is a slight technical point that we need to address here, which is the same technical point that arose in the previous lecture when we defined the direction of $\mathbf{A} \times \mathbf{B}$. We will choose $\mathbf{i}$, $\mathbf{j}$, and $\mathbf{k}$ in such a way that our coordinate system is a *right-handed coordinate system*. Here is the rule for doing that (the same as for choosing the direction of the cross product). Position your *right* hand so that your thumb is perpendicular to your fingers. Then put your thumb in the direction of $\mathbf{i}$ and your fingers in the direction of $\mathbf{j}$. Now there are two choices for the direction of $\mathbf{k}$ (remember, these three unit vectors are taken to be perpendicular). Our coordinate system will be right-handed if we choose $\mathbf{k}$ to be pointing out of the palm of our hand, see Fig. 2.1.

Note that it also follows, from the definition of the scalar product as well as the definition given for $\mathbf{i}$, $\mathbf{j}$, and $\mathbf{k}$, that:

> **Key point:**
>
> $$\mathbf{i} \cdot \mathbf{i} = 1, \, \mathbf{j} \cdot \mathbf{j} = 1, \, \mathbf{k} \cdot \mathbf{k} = 1. \qquad (2.2)$$

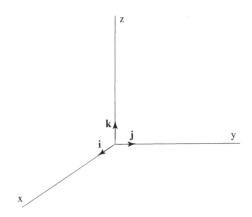

Fig. 2.1. The $\mathbf{i}$, $\mathbf{j}$, and $\mathbf{k}$ unit vectors in the direction of the x, y, and z axes, respectively.

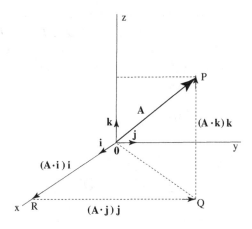

Fig. 2.2. The projection of **A** onto **i**, **j**, and **k**.

Expressing a Vector in Coordinates. Now consider an arbitrary vector **A** emanating from the origin of this coordinate system, as shown in Fig. 2.2.

From this figure we see geometrically (i.e. using the geometrical definition of vector addition given in Chapter 1) that **A** can be expressed as:

Key point:

$$\mathbf{A} = (\mathbf{A} \cdot \mathbf{i})\mathbf{i} + (\mathbf{A} \cdot \mathbf{j})\mathbf{j} + (\mathbf{A} \cdot \mathbf{k})\mathbf{k},$$

---

and the *numbers* $\mathbf{A} \cdot \mathbf{i}$, $\mathbf{A} \cdot \mathbf{j}$, and $\mathbf{A} \cdot \mathbf{k}$ are called the *components of* **A** *with respect to this coordinate system.* They are the x, y, and z coordinates of the tip of the vector **A**. For notational convenience we may write:

$$A_1 \equiv \mathbf{A} \cdot \mathbf{i}, \ A_2 \equiv \mathbf{A} \cdot \mathbf{j}, \ A_3 \equiv \mathbf{A} \cdot \mathbf{k},$$

or

$$A_x \equiv \mathbf{A} \cdot \mathbf{i}, \ A_y \equiv \mathbf{A} \cdot \mathbf{j}, \ A_z \equiv \mathbf{A} \cdot \mathbf{k}.$$

Using Coordinates to Compute the Magnitude of a Vector. We can also compute the length (or, magnitude) of **A** in terms of its components (and this is one place where coordinates are very useful). Refer to Fig. 2.2. Let P denote the point at the tip of the vector **A**, let Q denote its projection onto the plane defined by **i** and **j** (projection along the **k** direction), and let R denote the endpoint of $(\mathbf{A} \cdot \mathbf{i})\mathbf{i}$. Let OP denote the line segment from O to P, QP denote the line segment from Q to P, OQ denote the line segment from O to Q, OR denote the line segment from O to R, and RQ denote the line segment from R to Q. Let $\overline{OP}$, $\overline{QP}$, $\overline{OQ}$, $\overline{OR}$, and $\overline{RQ}$ denote the lengths of these respective segments. Then the length of $\mathbf{A}$ is $\overline{OP}$ and we will derive an expression for it in terms of the components of $\mathbf{A}$.

From the Pythagorean theorem we have:

$$\overline{OP}^2 = \overline{OQ}^2 + \overline{QP}^2, \tag{2.3}$$

and

$$\overline{OQ}^2 = \overline{OR}^2 + \overline{RQ}^2. \tag{2.4}$$

Substituting (2.4) into (2.3) gives:

$$\overline{OP}^2 = \overline{OR}^2 + \overline{RQ}^2 + \overline{QP}^2, \tag{2.5}$$

or,

$$|\mathbf{A}|^2 = A_1^2 + A_2^2 + A_3^2. \tag{2.6}$$

Therefore, we have:

> **Key point:**
>
> $$|\mathbf{A}| = \sqrt{A_1^2 + A_2^2 + A_3^2}. \tag{2.7}$$

Note that in passing from (2.6) to (2.7) a choice of sign was made in taking the square root. Why did we choose the positive sign? The choice of sign always needs to be taken into account when taking any root of a number. Taking it for granted will eventually cause some regret. (But in this case we used the general fact that "length" or "distance" is always non-negative.)

There is another issue here that is important, but often taken for granted. We chose a particular coordinate system to compute the length of a vector. However, one might imagine that "length" should be an intrinsic property of a vector and, therefore, be a number that is *independent* of the particular coordinate system with respect to which it is computed. In the realm of classical mechanics this is true (but this notion may require reconsideration when you learn about the theory of special relativity).

Vector Addition and Multiplication by Scalars in Coordinates. Vector addition in coordinates is easy. You just add the components. Let

$$\mathbf{A} = A_1\mathbf{i} + A_2\mathbf{j} + A_3\mathbf{k},$$
$$\mathbf{B} = B_1\mathbf{i} + B_2\mathbf{j} + B_3\mathbf{k}.$$

Then

Key point:

$$\mathbf{A} + \mathbf{B} = (A_1 + B_1)\mathbf{i} + (A_2 + B_2)\mathbf{j} + (A_3 + B_3)\mathbf{k}.$$

Multiplication of a vector by a scalar is also straightforward. Let a be a scalar. Then:

Key point:

$$a\mathbf{A} = aA_1\mathbf{i} + aA_2\mathbf{j} + aA_3\mathbf{k}.$$

Identification of a Vector with a 3-Tuple. Once our coordinate system is chosen, then the components of a vector (in the chosen coordinate system) uniquely define the vector. Hence, we could identify any vector with the 3-tuple consisting of the components of the vector, i.e.

$$(A_1, A_2, A_3).$$

Vector addition corresponds to adding 3-tuples as follows:

$$(A_1, A_2, A_3) + (B_1, B_2, B_3) = (A_1 + B_1, A_2 + B_2, A_3 + B_3),$$

and scalar multiplication of a vector corresponds to scalar multiplication of the 3-tuple as follows:

$$a(A_1, A_2, A_3) = (aA_1, aA_2, aA_3).$$

The Dot, or Scalar Product in Coordinates. Using (2.1) and (2.2), the dot product of vectors is easy to compute when the vectors are represented in these coordinates. Let

$$\mathbf{A} = A_1\mathbf{i} + A_2\mathbf{j} + A_3\mathbf{k},$$
$$\mathbf{B} = B_1\mathbf{i} + B_2\mathbf{j} + B_3\mathbf{k}.$$

Then

Key point:

$$\mathbf{A} \cdot \mathbf{B} = A_1 B_1 + A_2 B_2 + A_3 B_3. \qquad (2.8)$$

It also follows immediately by comparing (2.8) (with $\mathbf{A} = \mathbf{B}$) and (2.7) that:

Key point:

$$|\mathbf{A}| = \sqrt{\mathbf{A} \cdot \mathbf{A}}.$$

The Vector, or Cross Product in Coordinates. It is also easy to compute the cross-product of two vectors in coordinates. However, first we need to derive some properties of cross-products of the unit vectors.

First, it should be clear that:

Key point:

$$\mathbf{i} \times \mathbf{i} = 0, \ \mathbf{j} \times \mathbf{j} = 0, \ \mathbf{k} \times \mathbf{k} = 0, \qquad (2.9)$$

---

since the cross-product of two parallel vectors is zero.

Now consider $\mathbf{i} \times \mathbf{j}$. By definition this is:

$$\mathbf{i} \times \mathbf{j} = |\mathbf{i}| \, |\mathbf{j}| \, \sin \frac{\pi}{2} \mathbf{n} = \mathbf{n} = \mathbf{k},$$

The unit vector $\mathbf{n}$ is from our original definition of cross-product. The only issue here is why did we identify $\mathbf{n}$ with $\mathbf{k}$? This is because we have chosen our coordinate system according to the *right-hand rule* described above. Reasoning in the same way, we have:

Key point:

$$\mathbf{i} \times \mathbf{j} = \mathbf{k}, \quad \mathbf{j} \times \mathbf{i} = -\mathbf{k},$$
$$\mathbf{j} \times \mathbf{k} = \mathbf{i}, \quad \mathbf{k} \times \mathbf{j} = -\mathbf{i},$$
$$\mathbf{k} \times \mathbf{i} = \mathbf{j}, \quad \mathbf{i} \times \mathbf{k} = -\mathbf{j}. \qquad (2.10)$$

With (2.9) and (2.10) in hand, we have the means to compute the cross-product of two vectors. Let

$$\mathbf{A} = A_1\mathbf{i} + A_2\mathbf{j} + A_3\mathbf{k},$$

$$\mathbf{B} = B_1\mathbf{i} + B_2\mathbf{j} + B_3\mathbf{k},$$

and we wish to compute:

$$\mathbf{A} \times \mathbf{B} = (A_1\mathbf{i} + A_2\mathbf{j} + A_3\mathbf{k}) \times (B_1\mathbf{i} + B_2\mathbf{j} + B_3\mathbf{k})$$

$$= (A_1\mathbf{i} + A_2\mathbf{j} + A_3\mathbf{k}) \times (B_1\mathbf{i} + B_2\mathbf{j})$$
$$+ (A_1\mathbf{i} + A_2\mathbf{j} + A_3\mathbf{k}) \times B_3\mathbf{k},$$

$$= (A_1\mathbf{i} + A_2\mathbf{j} + A_3\mathbf{k}) \times B_1\mathbf{i}$$
$$+ (A_1\mathbf{i} + A_2\mathbf{j} + A_3\mathbf{k}) \times B_2\mathbf{j}$$
$$+ (A_1\mathbf{i} + A_2\mathbf{j} + A_3\mathbf{k}) \times B_3\mathbf{k},$$

$$= A_1\mathbf{i} \times B_1\mathbf{i} + A_2\mathbf{j} \times B_1\mathbf{i} + A_3\mathbf{k} \times B_1\mathbf{i}$$
$$+ A_1\mathbf{i} \times B_2\mathbf{j} + A_2\mathbf{j} \times B_2\mathbf{j} + A_3\mathbf{k} \times B_2\mathbf{j}$$
$$+ A_1\mathbf{i} \times B_3\mathbf{k} + A_2\mathbf{j} \times B_3\mathbf{k} + A_3\mathbf{k} \times B_3\mathbf{k},$$

$$= (A_2B_3 - A_3B_2)\mathbf{i} + (A_3B_1 - A_1B_3)\mathbf{j} + (A_1B_2 - A_2B_1)\mathbf{k}.$$

A useful "mnemonic" for remembering the expression for the cross-product, $\mathbf{A} \times \mathbf{B}$, in cartesian coordinates is through the determinant, expressed as follows:

Key point:

$$\mathbf{A} \times \mathbf{B} = \det \begin{vmatrix} \mathbf{i} & \mathbf{j} & \mathbf{k} \\ A_1 & A_2 & A_3 \\ B_1 & B_2 & B_3 \end{vmatrix}$$

$$= (A_2B_3 - A_3B_2)\mathbf{i} + (A_3B_1 - A_1B_3)\mathbf{j} + (A_1B_2 - A_2B_1)\mathbf{k}.$$

Kinematics–Derivatives and Integrals of Vectors

Suppose you have a vector, $\mathbf{A}$, that is a function of a scalar variable, t, i.e.

$$\mathbf{A} = \mathbf{A}(t).$$

What would we mean by "the derivative of $\mathbf{A}(t)$", and what types of properties would it satisfy? This is a key point for the subject of mechanics because we have already said that many of the quantities of interest will be expressed as vectors, and mechanics is the study of how those quantities change in time (which is the reason we chose to denote the scalar variable by t). Since the mathematical operation of differentiation is concerned with the instantaneous rate of change (of whatever we may be differentiating) it stands to reason that understanding the idea of the derivative of a vector will be important to us. Remember, if a vector "changes", it can change in two ways; direction and length, since these are the two characteristics that define a vector.

We will assume familiarity with elementary properties of differentiation and integration of functions of one variable.

Differentiation of Vectors. The derivative of a vector is defined in terms of the usual difference quotient that you are already familiar with (hopefully):

$$\frac{d\mathbf{A}}{dt}(t) = \lim_{\Delta t \to 0} \frac{\mathbf{A}(t + \Delta t) - \mathbf{A}(t)}{\Delta t}, \quad \text{provided the limit exists.} \quad (2.11)$$

However, in this course we are only going to differentiate and integrate vectors that are represented in some coordinate system, for now, the coordinate system will be that defined by the unit vectors $\mathbf{i}$, $\mathbf{j}$ and $\mathbf{k}$. So letting:

$$\mathbf{A}(t) = A_1(t)\mathbf{i} + A_2(t)\mathbf{j} + A_3(t)\mathbf{k}, \quad (2.12)$$

then the derivative of $\mathbf{A}(t)$ is given by:

> **Key point:**
> $$\frac{d\mathbf{A}}{dt}(t) = \frac{dA_1}{dt}(t)\mathbf{i} + \frac{dA_2}{dt}(t)\mathbf{j} + \frac{dA_3}{dt}(t)\mathbf{k}. \quad (2.13)$$

Is it obvious that (2.13) follows from the definition (2.11)? Probably not, it does require a proof, but we will not do that in this book. However, you could try to derive (2.13) from (2.11) yourself. A key fact is that the derivatives of $\mathbf{i}$, $\mathbf{j}$ and $\mathbf{k}$ are zero, i.e. they don't change in length or direction. This follows from their definition, but it is an important point to realize since later we will learn about *moving coordinate systems*, and in those the unit vectors defining the coordinate axes may change their direction (with respect to ...?).

Example 1. Let

$$\mathbf{A}(t) = e^{-t}\sin t\mathbf{i} + \ln t\mathbf{j} + \frac{t^3}{3}\mathbf{k},$$

then

$$\frac{d\mathbf{A}}{dt}(t) = (-e^{-t}\sin t + e^{-t}\cos t)\mathbf{i} + \frac{1}{t}\mathbf{j} + t^2\mathbf{k}.$$

So you see that when the vector is represented in cartesian coordinates, differentiation just means differentiating each of the components. Therefore you only need the knowledge of the standard calculus of one variable.

Higher order derivatives are computed in the same way. For example, the second derivative of $\mathbf{A}(t)$ is expressed as:

$$\frac{d^2\mathbf{A}}{dt^2}(t) = \frac{d^2A_1}{dt^2}(t)\mathbf{i} + \frac{d^2A_2}{dt^2}(t)\mathbf{j} + \frac{d^2A_3}{dt^2}(t)\mathbf{k}.$$

We compute the second derivative of the example given above.

Example 2. Let

$$\mathbf{A}(t) = e^{-t}\sin t\mathbf{i} + \ln t\mathbf{j} + \frac{t^3}{3}\mathbf{k},$$

then

$$\frac{d^2\mathbf{A}}{dt^2}(t) = (e^{-t}\sin t - e^{-t}\cos t - e^{-t}\cos t - e^{-t}\sin t)\mathbf{i} - \frac{1}{t^2}\mathbf{j} + 2t\mathbf{k}.$$

$$= -2e^{-t}\cos t\mathbf{i} - \frac{1}{t^2}\mathbf{j} + 2t\mathbf{k}$$

In calculus you learned how to differentiate products of functions. Similar rules exist for vector calculus. However, in vector calculus there are three types of products: the product of a vector function with a scalar function, the scalar product of two vectors, and the cross-product of two vectors. For a scalar function $a(t)$, and vector functions $\mathbf{A}(t)$, $\mathbf{B}(t)$, the rules for differentiating these products are given below.

> **Key point:**
>
> **"Product Rules" for Differentiating Products Involving Vectors**
>
> $$\frac{d}{dt}\left(a(t)\mathbf{A}(t)\right) = a(t)\frac{d\mathbf{A}}{dt}(t) + \frac{da}{dt}(t)\mathbf{A}(t),$$
>
> $$\frac{d}{dt}\left(\mathbf{A}(t)\cdot\mathbf{B}(t)\right) = \mathbf{A}(t)\cdot\frac{d\mathbf{B}}{dt}(t) + \frac{d\mathbf{A}}{dt}(t)\cdot\mathbf{B}(t),$$
>
> $$\frac{d}{dt}\left(\mathbf{A}(t)\times\mathbf{B}(t)\right) = \mathbf{A}(t)\times\frac{d\mathbf{B}}{dt}(t) + \frac{d\mathbf{A}}{dt}(t)\times\mathbf{B}(t).$$

Integration of Vectors. Integration of vector functions expressed in coordinates is developed in an analogous way. We have:

$$\int \mathbf{A}(t)dt = \int A_1(t)dt\,\mathbf{i} + \int A_2(t)dt\,\mathbf{j} + \int A_3(t)dt\,\mathbf{k} \qquad (2.14)$$

There is a corresponding "fundamental theorem of calculus" for integrals of vectors. If there exists a vector function $\mathbf{B}(t)$ such that

$$\mathbf{A}(t) = \frac{d\mathbf{B}}{dt}(t), \qquad (2.15)$$

then the indefinite integral of $\mathbf{A}(t)$ is given by:

$$\int \mathbf{A}(t)dt = \int \frac{d\mathbf{B}}{dt}(t)dt = \mathbf{B}(t) + \mathbf{c}, \qquad (2.16)$$

where $\mathbf{c}$ is a constant *vector*. The definite integral of $\mathbf{A}(t)$ is expressed as:

$$\int_a^b \mathbf{A}(t)dt = \int_a^b \frac{d\mathbf{B}}{dt}(t)dt = (\mathbf{B}(t) + \mathbf{c})\,|_a^b = \mathbf{B}(b) - \mathbf{B}(a). \qquad (2.17)$$

We give an example.

Example 3. Let

$$\mathbf{A}(t) = e^{-t}\sin t\,\mathbf{i} + \ln t\,\mathbf{j} + \frac{t^3}{3}\mathbf{k},$$

and we want to compute

$$\int \mathbf{A}(t)dt.$$

Now

$$\int e^{-t} \sin t \, dt = -\frac{e^{-t}}{2}(\sin t + \cos t) + c_1,$$

$$\int \ln t \, dt = t \ln t - t + c_2,$$

$$\int \frac{t^3}{3} \, dt = \frac{t^4}{12}.$$

Therefore

$$\int \mathbf{A}(t)dt = -\frac{e^{-t}}{2}(\sin t + \cos t)\mathbf{i} + (t \ln t - t)\mathbf{j} + \frac{t^4}{12}\mathbf{k} + c_1\mathbf{i} + c_2\mathbf{j} + c_3\mathbf{k}.$$

Kinematics–Space Curves, their Description and Derivatives

If a particle is in motion we can imagine it tracing out a curve in space. In order to give a mathematical description of this motion we need to develop some mathematical tools for describing a curve in three-dimensional space, or a *space curve*.

Space Curves, or Paths. Consider the following vector that depends on a scalar variable t ("time"):

$$\mathbf{r}(t) = x(t)\mathbf{i} + y(t)\mathbf{j} + z(t)\mathbf{k} \tag{2.18}$$

For each t this vector locates a point in space. As t varies, the tip of the vector traces out a curve. We say that (2.18) defines a *space curve*, see Fig. 2.3. We imagine that this curve is the *path* of a particle.

Velocity. We define the velocity of the particle at the point P (also called *instantaneous velocity*) as:

$$\mathbf{v}(t) = \lim_{\Delta t \to 0} \frac{\mathbf{r}(t + \Delta t) - \mathbf{r}(t)}{\Delta t}, \quad \text{if the limit exists.} \tag{2.19}$$

Geometrically, it is a vector tangent to the path at the point P. In coordinates, we have:

$$\mathbf{v}(t) = \frac{d\mathbf{r}}{dt}(t) = \frac{dx}{dt}(t)\mathbf{i} + \frac{dy}{dt}(t)\mathbf{j} + \frac{dz}{dt}(t)\mathbf{k}.$$

The magnitude of the velocity is called the *speed*, and is given by:

$$|\mathbf{v}(t)| = \left|\frac{d\mathbf{r}}{dt}(t)\right| = \sqrt{\left(\frac{dx}{dt}(t)\right)^2 + \left(\frac{dy}{dt}(t)\right)^2 + \left(\frac{dz}{dt}(t)\right)^2} \equiv \frac{ds}{dt}(t),$$

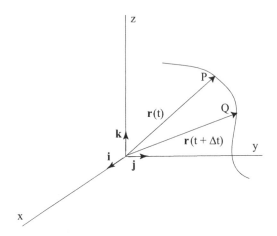

Fig. 2.3. Geometry of a space curve.

where s is the *arclength* along the curve as measured from some initial point P. More precisely, we imagine the *components* of the curve, $(x(t), y(t), z(t))$ as given, then this is a differential equation whose solution is the arclength along the curve (but you must choose a starting and ending point in order to do the integral).

Acceleration. Acceleration of the particle at the point P (or, sometimes called *instantaneous acceleration*) is the derivative of the velocity at the point P, i.e.

$$\mathbf{a}(t) = \frac{d\mathbf{v}}{dt}(t) = \lim_{\Delta t \to 0} \frac{\mathbf{v}(t + \Delta t) - \mathbf{v}(t)}{\Delta t}, \quad \text{if the limit exists.}$$

In cartesian coordinates this is expressed as:

$$\mathbf{a}(t) = \frac{d\mathbf{v}}{dt}(t) = \frac{d^2 x}{dt^2}(t)\mathbf{i} + \frac{d^2 y}{dt^2}(t)\mathbf{j} + \frac{d^2 z}{dt^2}(t)\mathbf{k},$$

and the magnitude of the acceleration is given by:

$$|\mathbf{a}(t)| = \left| \frac{d\mathbf{v}}{dt}(t) \right| = \sqrt{\left(\frac{d^2 x}{dt^2}(t) \right)^2 + \left(\frac{d^2 y}{dt^2}(t) \right)^2 + \left(\frac{d^2 z}{dt^2}(t) \right)^2}.$$

Problem Set 2

1. Suppose $\mathbf{A}$, $\mathbf{B}$, $\mathbf{C}$,and $\mathbf{D}$ are vectors and a is a scalar. Using the definition of the cross, or vector product, given in class, *and expressing the*

vectors in rectangular coordinates, show that the following properties hold.

(a) $\mathbf{A} \times \mathbf{B} = -\mathbf{B} \times \mathbf{A}$,   Commutative Law for Cross-Products Fails,

(b) $(\mathbf{A} + \mathbf{B}) \times (\mathbf{C} + \mathbf{D}) = (\mathbf{A} \times \mathbf{C}) + (\mathbf{B} \times \mathbf{C}) + (\mathbf{A} \times \mathbf{D}) + (\mathbf{B} \times \mathbf{D})$, Distributive Law,

(c) $a(\mathbf{A} \times \mathbf{B}) = (a\mathbf{A}) \times \mathbf{B} = \mathbf{A} \times (a\mathbf{B}) = (\mathbf{A} \times \mathbf{B})a$.

2. Let $\mathbf{A} = 4\mathbf{i} + 3\mathbf{j} + 2\mathbf{k}$ and $\mathbf{B} = -\mathbf{i} + 7\mathbf{j} - 3\mathbf{k}$. Compute $\mathbf{A} \times \mathbf{B}$ and $\mathbf{A} \cdot \mathbf{B}$.

3. Let $\mathbf{A} = 2\mathbf{i} + 4\mathbf{j} + 2\mathbf{k}$ and $\mathbf{B} = -4\mathbf{i} - 8\mathbf{j} - 4\mathbf{k}$. Determine $\mathbf{A} \times \mathbf{B}$ *without doing any computations*.

4. Consider each of the expressions below. Which expressions "make sense" (i.e. have mathematical validity in terms of the properties of vectors). For those that do, use $\mathbf{a} = \mathbf{i} - \mathbf{j}$, $\mathbf{b} = \mathbf{i}$, $\mathbf{c} = \mathbf{j}$, and $\mathbf{d} = \mathbf{i} + \mathbf{j}$ to evaluate the expression.[1]

(a) $\frac{\mathbf{a} \cdot \mathbf{b}}{\mathbf{c} \cdot \mathbf{d}}$.

(b) $\mathbf{a} \cdot \mathbf{b} \cdot \mathbf{c}$,

(c) $\frac{\mathbf{a} \cdot \mathbf{b}\,\mathbf{d}}{\mathbf{b} \cdot \mathbf{c}}$.

(d) $\frac{\mathbf{a} \cdot \mathbf{d}\,\mathbf{b} \cdot \mathbf{c}}{\mathbf{d}}$.

(e) $|\mathbf{a} \cdot \mathbf{b}\,\mathbf{d}|$,

(f) $\mathbf{a}\,|\mathbf{c} - \mathbf{d}|$,

(g) $\mathbf{a} + \mathbf{b} \cdot \mathbf{c}$,

(h) $\mathbf{a} \times \mathbf{b}$,

(i) $\mathbf{a} \times \mathbf{b} \times \mathbf{c}$,

(j) $\mathbf{a} \times \mathbf{b} + \mathbf{a} \cdot \mathbf{b}$,

(k) $\mathbf{a} \times \mathbf{b} \cdot \mathbf{d}$,

(l) $\mathbf{a} \times (\mathbf{b} \cdot \mathbf{c})$.

5. Consider the vectors

$$\mathbf{A} = A_1\mathbf{i} + A_2\mathbf{j} + A_3\mathbf{k}, \quad \mathbf{B} = B_1\mathbf{i} + B_2\mathbf{j} + B_3\mathbf{k},$$

both emanating from the origin of a cartesian coordinate system. Let the tip of the vector $\mathbf{A}$ be denoted by P and the tip of the vector $\mathbf{B}$ be denoted by Q. In other words, with respect to the cartesian coordinate system, P is located at the point (A_1, A_2, A_3) and Q is located at the point (B_1, B_2, B_3). Determine the vector starting at P and ending at Q, and compute its magnitude. See the figure below.

---

[1]The point of this exercise is to emphasize basic algebraic properties of vectors. The dot product of two vectors is a scalar, the cross-product of two vectors is a vector, and dividing by a vector is not defined.

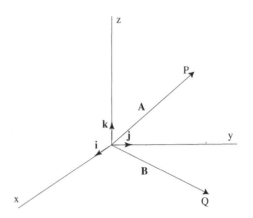

6. Let $\mathbf{A}(t) = A_1(t)\mathbf{i} + A_2(t)\mathbf{j} + A_3(t)\mathbf{k}$, $\mathbf{B}(t) = B_1(t)\mathbf{i} + B_2(t)\mathbf{j} + B_3(t)\mathbf{k}$, and $a(t)$ denote a scalar valued function. Prove the following properties for different types of products.

   (a) $\frac{d}{dt}\left(a(t)\mathbf{A}(t)\right) = a(t)\frac{d\mathbf{A}}{dt}(t) + \frac{da}{dt}(t)\mathbf{A}(t)$,

   (b) $\frac{d}{dt}\left(\mathbf{A}(t)\cdot\mathbf{B}(t)\right) = \mathbf{A}(t)\cdot\frac{d\mathbf{B}}{dt}(t) + \frac{d\mathbf{A}}{dt}(t)\cdot\mathbf{B}(t)$,

   (c) $\frac{d}{dt}\left(\mathbf{A}(t)\times\mathbf{B}(t)\right) = \mathbf{A}(t)\times\frac{d\mathbf{B}}{dt}(t) + \frac{d\mathbf{A}}{dt}(t)\times\mathbf{B}(t)$.

7. If $\mathbf{A}(t) = 4(t-1)\mathbf{i} - (2t+3)\mathbf{j} + 6t^2\mathbf{k}$, then compute:[2]

   (a) $\displaystyle\int_2^3 \mathbf{A}(t)dt$,

   (b) $\displaystyle\int_1^2 (t\mathbf{i} - 2\mathbf{k})\cdot\mathbf{A}(t)dt$.

8. Suppose a particle moves along a space curve defined by:

$$x(t) = e^{-t}\cos t,\ y(t) = e^{-t}\sin t,\ z(t) = e^{-t}.$$

   Find the magnitude of the velocity and acceleration at any time t.

9. The position vector of a particle at any time t is given by:

$$\mathbf{r} = a\cos\omega t\mathbf{i} + a\sin\omega t\mathbf{j} + bt^2\mathbf{k}.$$

   where a and b are scalars. Show that the speed of the particle increases with time, but the magnitude of the acceleration is constant. Describe the motion of the particle geometrically.

---

[2]This problem can be a bit tricky. It is important to understand the nature of each integrand.

Chapter 3

Kinematics — Space Curves, their Description and Derivatives, Circular Motion, and Line Integrals

A "Moving Coordinate System" Along a Space Curve. We will now construct a coordinate system (i.e. a set of three, orthogonal unit vectors) *at each point* of a space curve. This can be viewed in two ways. If we imagine a particle moving along the curve then our coordinate system could be viewed as a *moving coordinate system*, moving with the particle (with the particle always at the origin of the coordinate system). Alternatively, we could view the curve as being fixed in space and the coordinate system at each point could be used to describe the shape of the curve. Both points of view occur in many applications. For example, in biology DNA molecules are modeled as curves with a complicated folded structure. In physics and chemistry polymers are also modeled as curves with a complicated spatial structure. In pure mathematics the subject of knot theory is concerned with describing the structure of closed curves in three dimensions.

We begin by considering a curve, which we refer to as C, that is traced out by the vector function $\mathbf{r} = \mathbf{r}(t)$, as t varies through all possible values. We consider a particular point on that curve, called P, see Fig. 3.1.

Notation: Make sure you distinguish vectors (boldface type) from scalars (regular type) in the following. Also, all quantities are defined at a specific point on the curve, i.e. they are functions of t. Remember this, because

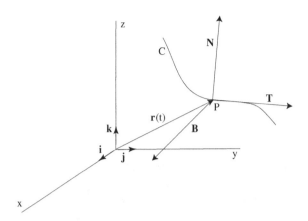

Fig. 3.1. Geometry of the coordinate system associated with a space curve.

for the sake of less cumbersome notation we are going to omit the explicit dependence on t.

First, we define the *unit tangent vector* $\mathbf{T}$, tangent to C at P. This is defined as:

$$\mathbf{T} = \frac{d\mathbf{r}}{ds}, \tag{3.1}$$

where s is arclength measured from some initial point on the curve (which is arbitrary). So where does this come from, and why is it of unit length?

First, the velocity, $\mathbf{v}(t) \equiv \frac{d\mathbf{r}}{dt}(t)$ is, by definition, tangent to C. Now recall the definition of arclength given earlier:

Key point:

$$|\mathbf{v}(t)| = \left|\frac{d\mathbf{r}}{dt}(t)\right| = \sqrt{\left(\frac{dx}{dt}(t)\right)^2 + \left(\frac{dy}{dt}(t)\right)^2 + \left(\frac{dz}{dt}(t)\right)^2} \equiv \frac{ds}{dt}. \tag{3.2}$$

A good candidate for a unit tangent vector to C at P would be

$$\frac{\mathbf{v}(t)}{|\mathbf{v}(t)|}. \tag{3.3}$$

Now if you remember the chain rule from calculus, it follows from (3.2) and the definition of velocity, that

> **Key point:**
> $$\frac{\mathbf{v}(t)}{|\mathbf{v}(t)|} = \frac{d\mathbf{r}}{dt}\frac{dt}{ds}(t) = \frac{d\mathbf{r}}{ds} = \mathbf{T}, \tag{3.4}$$

which is the same as (3.1).

Technical Point: There is an issue here that we have side-stepped. What happens if $\mathbf{v}(t) = 0$? Dividing by zero is (usually) not allowed. In this book we will not explicitly deal with this issue. When you learn more about curves and surfaces you will learn how to understand this point.

Next we define the *unit principal normal* $\mathbf{N}$ at P. This is defined as:

> **Key point:**
> $$\mathbf{N} \equiv \frac{d\mathbf{T}}{ds} \bigg/ \left|\frac{d\mathbf{T}}{ds}\right|. \tag{3.5}$$

Clearly, $\mathbf{N}$ is of unit length (at least it should be "clear", or else you have missed something really fundamental and should go back and find out what it is), but it should also be *perpendicular* to $\mathbf{T}$. This is (probably) not obvious, so we will prove it. We have made the argument (by intimidation) that $\mathbf{T}$ is of unit length. Therefore,

$$\mathbf{T} \cdot \mathbf{T} = 1. \tag{3.6}$$

Now let us differentiate this expression with respect to s:

$$\mathbf{T} \cdot \frac{d\mathbf{T}}{ds} + \frac{d\mathbf{T}}{ds} \cdot \mathbf{T} = 2\mathbf{T} \cdot \frac{d\mathbf{T}}{ds} = 0, \tag{3.7}$$

from which it is immediate that:

$$\mathbf{T} \cdot \frac{d\mathbf{T}}{ds} = 0. \tag{3.8}$$

From this expression, and the definition of $\mathbf{N}$ given in (3.5), it follows that $\mathbf{N} \cdot \mathbf{T} = 0$. At this point, we introduce some notation and terminology. It is standard to define:

> **Key point:**
> $$\left|\frac{d\mathbf{T}}{ds}\right| \equiv \kappa, \tag{3.9}$$

to be the *curvature of C at P*. Moreover,

Key point:

$$R = \frac{1}{\kappa} \tag{3.10}$$

is defined to be the *radius of curvature of C at P*. With these definitions we can write:

$$\mathbf{N} \equiv R\frac{d\mathbf{T}}{ds}. \tag{3.11}$$

We complete the construction of our coordinate system by defining the *unit binormal* $\mathbf{B}$ *to C at P*. To do this, we desire to construct a third unit vector that is perpendicular to the two that we have already constructed. How would we do this? You should already have a good idea. We merely need to take the cross-product of the two unit vectors that we have already constructed:

Key point:

$$\mathbf{B} = \mathbf{T} \times \mathbf{N}. \tag{3.12}$$

Technical Point: Why did we take $\mathbf{B} = \mathbf{T} \times \mathbf{N}$, rather than $\mathbf{B} = \mathbf{N} \times \mathbf{T}$? Because we wanted a right-handed coordinate system. Go back and convince yourself that this is the type of coordinate system we have constructed.

It follows from our construction (see (3.1) and (3.4)) that the velocity of a particle on this curve is given by:

Key point:

$$\mathbf{v}(t) = v\mathbf{T}, \quad \text{where} \quad v \equiv |\mathbf{v}|. \tag{3.13}$$

What is the *acceleration* of a particle on this curve? We can compute this by differentiating (3.13):

$$\mathbf{a} = \frac{d\mathbf{v}}{dt} = \frac{d}{dt}(v\mathbf{T}) = \frac{dv}{dt}\mathbf{T} + v\frac{d\mathbf{T}}{dt}. \tag{3.14}$$

Now

$$\frac{d\mathbf{T}}{dt} = \frac{d\mathbf{T}}{ds}\frac{ds}{dt}. \tag{3.15}$$

From (3.2) it follows that

$$\frac{ds}{dt} = v, \tag{3.16}$$

and from (3.5) and (3.9) it follows that:

$$\frac{d\mathbf{T}}{ds} = \kappa\mathbf{N}. \tag{3.17}$$

Therefore

$$\frac{d\mathbf{T}}{dt} = \kappa\mathbf{N}\frac{ds}{dt} = \kappa v\mathbf{N} = \frac{v\mathbf{N}}{R}. \tag{3.18}$$

Putting this all together gives the following expression for the acceleration:

Key point:

$$\mathbf{a} = \frac{dv}{dt}\mathbf{T} + v\left(\frac{v\mathbf{N}}{R}\right) = \frac{dv}{dt}\mathbf{T} + \frac{v^2}{R}\mathbf{N}, \tag{3.19}$$

---

where the first and second terms on the right-hand side are called the *tangential acceleration* and *normal or centripetal acceleration*, respectively.[1]

Motion Constrained to a Circle. Now let us consider a very simple case of motion along a curve in space that arises in a number of applications: motion on a circle in the plane. This is an example of *constrained motion*, the particle is forced to move on the circle.

We consider a circle, C, of radius R, with center at the origin 0, and a particle P moving along the circle. If s denotes arclength measured along the circle (from some point A), then the angle subtended by the particle is given through the relation $s = R\theta$, see Fig. 3.2.

---

[1]
- For a space curve, $\mathbf{r}(t)$ (which we can think of as the position vector of a particle) we have defined an orthonormal coordinate system *at each point on the curve*, with the vectors $\mathbf{T}$, $\mathbf{N}$, and $\mathbf{B}$. These three vectors (generally) vary from point to point on this curve, so we say that $\mathbf{T}$, $\mathbf{N}$, and $\mathbf{B}$ define a *moving coordinate system*.
- The space curve can be parametrized by t or s, and the relation between these parameters is given by (3.2). This is a point that often gives students difficulties, at first. So think about what (3.2) means.
- The velocity and the acceleration of the particle located by this position vector $\mathbf{r}(t)$ can be described at each point in space defined by $\mathbf{r}(t)$ in terms of the (moving) coordinate system defined by $\mathbf{T}$, $\mathbf{N}$, and $\mathbf{B}$.

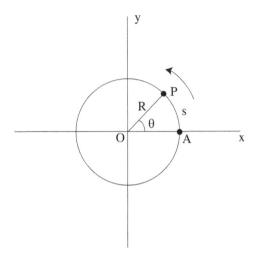

Fig. 3.2. Motion of a particle constrained to a circle.

The magnitudes of the tangential velocity and acceleration are given, respectively, by:

$$v = \frac{ds}{dt} = R\frac{d\theta}{dt} = R\omega, \tag{3.20}$$

$$\frac{dv}{dt} = \frac{d^2s}{dt^2} = R\frac{d^2\theta}{dt^2} = R\alpha, \tag{3.21}$$

where $\omega \equiv \frac{d\theta}{dt}$ is called the *angular speed* and $\alpha \equiv \frac{d^2\theta}{dt^2}$ is called the *angular acceleration.* The *normal or centripetal acceleration,* as seen from (3.19) is given by $\frac{v^2}{R} = \omega^2 R$.

Polar Coordinates, and a Rotating Coordinate System. Let (r, θ) denote the polar coordinates describing the position of a particle. Let $\mathbf{r}_1$ denote a unit vector in the direction of the position vector $\mathbf{r}$, and let $\boldsymbol{\theta}_1$ denote a unit vector perpendicular to $\mathbf{r}$, and in the direction of increasing θ, see Fig. 3.3. First, we want to derive expressions for $\mathbf{r}_1$ and $\boldsymbol{\theta}_1$ in terms of $\mathbf{i}$ and $\mathbf{j}$. The resulting equations will tell us how to transform vectors from one coordinate system to another (a subject that we will return to later). From Fig. 3.3, it is easy to see that the position vector $\mathbf{r}$ is given by:

$$\mathbf{r} = r\cos\theta\mathbf{i} + r\sin\theta\mathbf{j}. \tag{3.22}$$

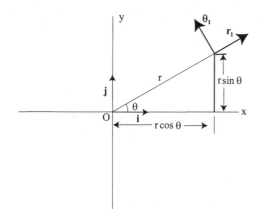

Fig. 3.3. The relationship between polar coordinates and Cartesian coordinates.

Therefore we have:

> **Key point:**
>
> $$r_1 = \frac{r}{|r|} = \cos\theta i + \sin\theta j. \tag{3.23}$$

It is also easy to verify that:

> **Key point:**
>
> $$\theta_1 = -\sin\theta i + \cos\theta j. \tag{3.24}$$

Moreover, one can solve (3.23) and (3.24) simultaneously for i and j as functions of r_1 and θ_1:

> **Key point:**
>
> $$i = \cos\theta r_1 - \sin\theta \theta_1, \tag{3.25}$$
>
> $$j = \sin\theta r_1 + \cos\theta \theta_1. \tag{3.26}$$

Now there is a big difference between i, j and r_1, θ_1. As the particle moves, i, j remain fixed in space with unit length (i.e. their derivatives with respect to t are zero), but r_1, θ_1 move with the particle. We want to determine how they move by computing their derivatives with respect to t. We begin with r_1. Differentiating (3.23) with respect to t gives:

Key point:

$$\dot{\mathbf{r}}_1 = -\sin\theta\,\dot{\theta}\mathbf{i} + \cos\theta\,\dot{\theta}\mathbf{j},$$

$$= \dot{\theta}\left(-\sin\theta\mathbf{i} + \cos\theta\mathbf{j}\right),$$

$$= \dot{\theta}\boldsymbol{\theta}_1, \qquad\qquad \textbf{using (3.24).} \qquad (3.27)$$

Next we compute the derivative of $\boldsymbol{\theta}_1$ with respect to t by differentiating (3.24):

Key point:

$$\dot{\boldsymbol{\theta}}_1 = -\cos\theta\,\dot{\theta}\mathbf{i} - \sin\theta\,\dot{\theta}\mathbf{j},$$

$$= -\dot{\theta}\left(\cos\theta\mathbf{i} + \sin\theta\mathbf{j}\right),$$

$$= -\dot{\theta}\mathbf{r}_1, \qquad\qquad \textbf{using (3.23).} \qquad (3.28)$$

Now we can compute the velocity of the particle in the coordinate system defined by $\mathbf{r}_1, \boldsymbol{\theta}_1$, where the position vector of the particle is $\mathbf{r} = r\mathbf{r}_1$:

Key point:

$$\mathbf{v} = \frac{d\mathbf{r}}{dt} = \frac{dr}{dt}\mathbf{r}_1 + r\frac{d\mathbf{r}_1}{dt} = \dot{r}\mathbf{r}_1 + r\dot{\mathbf{r}}_1 = \dot{r}\mathbf{r}_1 + r\dot{\theta}\boldsymbol{\theta}_1. \qquad (3.29)$$

Next we compute the acceleration of the particle in this coordinate system:

Key point:

$$\mathbf{a} = \frac{d\mathbf{v}}{dt} = \frac{d}{dt}\left(\dot{r}\mathbf{r}_1 + r\dot{\theta}\boldsymbol{\theta}_1\right),$$

$$= \ddot{r}\mathbf{r}_1 + \dot{r}\dot{\mathbf{r}}_1 + \dot{r}\dot{\theta}\boldsymbol{\theta}_1 + r\ddot{\theta}\boldsymbol{\theta}_1 + r\dot{\theta}\dot{\boldsymbol{\theta}}_1,$$

$$= \ddot{r}\mathbf{r}_1 + \dot{r}(\dot{\theta}\boldsymbol{\theta}_1) + \dot{r}\dot{\theta}\boldsymbol{\theta}_1 + r\ddot{\theta}\boldsymbol{\theta}_1 + (r\dot{\theta})(-\dot{\theta}\mathbf{r}_1),$$

$$= (\ddot{r} - r\dot{\theta}^2)\mathbf{r}_1 + (r\ddot{\theta} + 2\dot{r}\dot{\theta})\boldsymbol{\theta}_1. \qquad (3.30)$$

These formulae will be very useful later on.[2]

---

[2]The vectors $\mathbf{i}$ and $\mathbf{j}$ are of unit length, and their direction is fixed in space. Therefore their derivative with respect to time is zero. The vectors $\mathbf{r}_1$ and $\boldsymbol{\theta}_1$ are also of unit length, but their directions in space change. Therefore their derivatives with respect to time are nonzero, and we have computed these derivatives above. The derivatives of each vector can either be represented in the coordinate system defined by $\mathbf{i} - \mathbf{j}$, or the coordinate system defined by $\mathbf{r}_1 - \boldsymbol{\theta}_1$.

Kinematics–Line Integrals, and "Independence of Path"

Consider the cartesian coordinate system that we have developed and denote the coordinates of any point in space with respect to that coordinate system by (x, y, z). Suppose at each point of space we denote a vector, $\mathbf{A} = \mathbf{A}(x, y, z)$. Then we can view $\mathbf{A} = \mathbf{A}(x, y, z)$ as a *vector valued function* of the three variables (x, y, z). Sometimes such functions are called *vector fields*. Vector fields arise everywhere. For example, the mass of the Earth exerts a force at each point in space (the Earth's gravitational field) and in a flowing fluid each point of the fluid has a velocity (the velocity field of the fluid).

Key point: Be aware of the difference between a vector valued function of a single variable (e.g. a position vector that is a function of time), and a vector valued function of a vector variable (e.g. the velocity of a fluid defined at each point in space). The former should be familiar by now, the latter will become more familiar.

As a particle moves through space one may want to "add up" (or, more mathematically, "integrate") the effect a particular type of field has on the particle during the course of its motion. This is where the notion of *line, or path integral arises*.

Let

$$\mathbf{r}(t) = x(t)\mathbf{i} + y(t)\mathbf{j} + z(t)\mathbf{k}, \tag{3.31}$$

denote the position vector of a particle. Let $P_1 \equiv (x(t_1), y(t_1), z(t_1))$ and $P_2 \equiv (x(t_2), y(t_2), z(t_2))$ denote points on the path of the particle. We refer to the segment of the path of the particle between P_1 and P_2 as C. Let $\mathbf{A}(x, y, z) = A_1(x, y, z)\mathbf{i} + A_2(x, y, z)\mathbf{j} + A_3(x, y, z)\mathbf{k}$ denote a vector field, i.e. a vector defined at each point (x, y, z). The integral of the tangential component of $\mathbf{A}(x, y, z)$ along C from P_1 to P_2 is written as:

$$\int_{P_1}^{P_2} \mathbf{A} \cdot d\mathbf{r} = \int_C \mathbf{A} \cdot d\mathbf{r} = \int_C A_1 dx + A_2 dy + A_3 dz, \tag{3.32}$$

as an example of a line integral. If C is a closed curve it is usually written as:

$$\oint_C \mathbf{A} \cdot d\mathbf{r} = \oint_C A_1 dx + A_2 dy + A_3 dz, \tag{3.33}$$

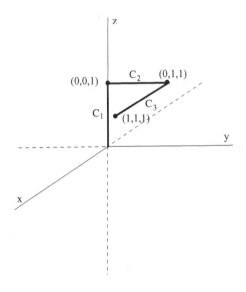

Fig. 3.4. Geometry of the path for computing the line integral.

It probably seems quite reasonable that the value of the line integral depends on the path, and generally it does.

Example 4. Consider the vector field

$$\mathbf{A} = (3x^2 - 6yz)\mathbf{i} + (2y + 3xz)\mathbf{j} + (1 - 4xyz^2)\mathbf{k}.$$

We will evaluate $\int_C \mathbf{A} \cdot d\mathbf{r}$ from $(0,0,0)$ to $(1,1,1)$ along the straight line from $(0,0,0)$ to $(0,0,1)$, then from $(0,0,1)$ to $(0,1,1)$, and, finally, from $(0,1,1)$, then to $(1,1,1)$. Figure 3.4 may help visualize the path.

First, we have:

$$\int_C \mathbf{A} \cdot d\mathbf{r} = \int_C \left((3x^2 - 6yz)\mathbf{i} + (2y + 3xz)\mathbf{j} \right.$$

$$\left. + (1 - 4xyz^2)\mathbf{k} \right) \cdot (dx\mathbf{i} + dy\mathbf{j} + dz\mathbf{k}),$$

$$= \int_C (3x^2 - 6yz)dx + (2y + 3xz)dy + (1 - 4xyz^2)dz.$$

$$(3.34)$$

From considering this expression, and the statement of the problem to be solved (i.e. compute this integral along a specified path between two points on the path) you should see one of the main issues concerning the computation of line integrals. There is an extra, essential, step involved.

We must develop a *parametrization of the path* between the two points for which the line integral is to be computed. In this case, we will give that parametrization explicitly. The path will consist of three segments:

1. C_1: the straight line along the z-axis from $(0,0,0)$ to $(0,0,1)$.
2. C_2: the straight line in the direction of the y-axis from $(0,0,1)$ to $(0,1,1)$.
3. C_3: the straight line in the direction of the x-axis from $(0,1,1)$ to $(1,1,1)$.

See Fig. 3.4.

Along the straight line from $(0,0,0)$ to $(0,0,1)$, $x = y = 0, dx = dy = 0$ and z varies from 0 to 1. Then the integral over this part of the path is:

$$\int_{z=0}^{z=1} (3(0)^2 - 6(0)z)d(0) + (2(0) + 3(0)z)d(0) + (1 - 4(0)(0)z^2)dz$$

$$= \int_{z=0}^{z=1} dz = 1.$$

Along the straight line from $(0,0,1)$ to $(0,1,1)$, $x = 0, z = 1, dx = 0, dz = 0$, and y varies from 0 to 1. Then the integral over this part of the path is:

$$\int_{y=0}^{y=1} (3(0)^2 - 6y)d(0) + (2y + 3(0))dy + (1 - 4(0)y)d(0) = \int_{y=0}^{y=1} 2ydy = 1.$$

Along the straight line from $(0,1,1)$ to $(1,1,1)$, $y = 1$, $z = 1$, $dy = 0$, $dz = 0$, and x varies from 0 to 1. Then the integral over this part of the path is:

$$\int_{x=0}^{x=1} (3x^2 - 6)dx + (2y + 3x)d(0) + (1 - 4x)d(0)$$

$$= \int_{x=0}^{x=1} (3x^2 - 6)dx = (x^3 - 6x)\Big|_0^1 = -5.$$

Adding up the contributions from each segment gives.

$$\int_C \mathbf{A} \cdot \mathbf{dr} = 1 + 1 - 5 = -3.$$

The "Art" of Computing Line Integrals

In order to compute (3.32) we need to parametrize the (one-dimensional) path connecting P_1 to P_2. In some problems a parametrization of the path may be specified (like in the example above). In others, it may be up to us to construct a parametrization. A clever choice of parametrization might make the computation of the line integral very easy.

We will revisit this example in the problems where we will compute the line integral along two different paths between $(0,0,0)$ and $(1,1,1)$, and we will see that we get a different answer for each path. This probably seems quite reasonable. However, in mechanics there are certain important situations when the line integral of *certain quantities* is independent of the path. This naturally gives rise to the following question. Is there something particular about the nature of a vector field $\mathbf{A}$ that would make its line integral between points P_1 and P_2 independent of the path taken from P_1 to P_2? The answer is "yes", but to tell you what the answer is we need to introduce some definitions.

First, we need to discuss a new notion of the derivative of a scalar valued function of more than one variable — the "partial derivative".

Partial Derivatives of Scalar Valued Functions of Two, or More Variables. We have seen that functions of more than one variable (e.g. functions of the coordinates of space) typically arise in mechanics. We will want to quantify how these functions change as the variables change. This leads to the notion of differentiation of functions of more than one variable. Over the course of your studies in mathematics you will learn that there are numerous ways that the differentiation operation can be carried out for high dimensional functions depending on many variables. Fundamentally, differentiation, in whatever context, is always related to the same idea — understanding the behaviour of a function *in a neighbourhood of a point* in terms of a linear approximation of the function *at the point.*

The notion of differentiation that we will need in this book is the notion of *partial differentiation*, which we will now describe. For our purposes we will be concerned with scalar valued function of two or three variables, e.g.

$$\phi(x, y, z) = xy \sin z, \qquad (3.35)$$

or

$$\psi(x, y) = \frac{x}{y}. \qquad (3.36)$$

In a sense you already know how to compute partial derivatives since the idea of partial differentiation reduces the differentiation problem for functions of more than one variable to the familiar situation of differentiation of a function of a single variable. For example, the partial derivative of (3.35) with respect to any *one* of the variables x, y, or z, let us say y, means that we view all variables *except* the one that we are differentiating with respect to as *fixed, or constant*. In this way we have a function of a single variable, y, and we differentiate the resulting function exactly as we differentiate functions of a single variable, e.g.

$$\frac{\partial \phi}{\partial y} = x \sin z, \tag{3.37}$$

where we use a different symbol for partial differentiation — ∂ rather than d.

We can also express the partial derivative of a *scalar valued function* of two or more variables in terms of the usual difference quotient. For example:

$$\frac{\partial \phi}{\partial y}(x, y, z) = \lim_{h \to 0} \frac{\phi(x, y + h, z) - \phi(x, y, z)}{h}, \quad \text{if the limit exists.} \tag{3.38}$$

For completeness, we give the remaining partial derivatives of $\phi(x, y, z)$:

$$\frac{\partial \phi}{\partial x} = y \sin z, \tag{3.39}$$

and

$$\frac{\partial \phi}{\partial z} = xy \cos z. \tag{3.40}$$

We also give the partial derivatives of $\psi(x, y)$:

$$\frac{\partial \psi}{\partial x} = \frac{1}{y}, \tag{3.41}$$

and

$$\frac{\partial \psi}{\partial y} = -\frac{x}{y^2}. \tag{3.42}$$

Gradient of a Scalar Valued Function (of a vector variable). Let $\phi(x, y, z)$ denote a scalar valued function of (x, y, z). The *gradient of ϕ, denoted $\nabla \phi$* is a *vector* whose components in cartesian coordinates are given by:

$$\nabla \phi = \left(\frac{\partial \phi}{\partial x}, \frac{\partial \phi}{\partial y}, \frac{\partial \phi}{\partial z} \right).$$

Curl of a Vector Valued Function (of a vector variable). Let $\mathbf{A} = \mathbf{A}(x, y, z)$ be a vector with components $(A_1(x, y, z), A_2(x, y, z), A_3(x, y, z))$. Then the *curl of $\mathbf{A} = \mathbf{A}(x, y, z)$, denoted $\nabla \times \mathbf{A}$*, is a *vector* whose components are given by:

$$\nabla \times \mathbf{A} = \left(\frac{\partial A_3}{\partial y} - \frac{\partial A_2}{\partial z}, \frac{\partial A_1}{\partial z} - \frac{\partial A_3}{\partial x}, \frac{\partial A_2}{\partial x} - \frac{\partial A_1}{\partial y} \right).$$

You will learn more about these notions when you study vector analysis. But it is always good to see the same concepts from different points of view and they arise very naturally in the setting of mechanics, as we will see in the coming chapters.

With these definitions in hand we can now state the conditions under which the line integral between two points is independent of the path taken between the point.

The line integral (3.32) will have a value that is *independent of the path joining* P_1 *and* P_2 if and only if:

$$\mathbf{A} = \nabla \phi, \tag{3.43}$$

for some scalar valued function $\phi = \phi(x, y, z)$ or, equivalently:

$$\nabla \times \mathbf{A} = 0. \tag{3.44}$$

First, we show that if $\mathbf{A}$ has the form of (3.43), then the line integral depends only on the endpoints of the path taken. Denote the coordinates of P_1 and P_2 by (x_1, y_1, z_1) and (x_2, y_2, z_2), respectively. Then

$$\int_{P_1}^{P_2} \mathbf{A} \cdot d\mathbf{r} = \int_{P_1}^{P_2} \nabla \phi \cdot d\mathbf{r} = \int_{P_1}^{P_2} \frac{\partial \phi}{\partial x} dx + \frac{\partial \phi}{\partial y} dy + \frac{\partial \phi}{\partial z} dz$$

$$= \int_{P_1}^{P_2} d\phi = \phi(x_2, y_2, z_2) - \phi(x_1, y_1, z_1).$$

So the value of this integral depends only on the endpoints.

Now it is not obvious that if $\mathbf{A}$ satisfies (3.44), then it can be written as the gradient of a scalar valued function. You may prove this in other topics that you may study. However, for now, it provides a computable criterion to determine of the line integral of a vector field along a path depends only on the endpoints.

Key point: The line integral of a vector field $\mathbf{A}(x, y, z)$ between two points, P_1 and P_2, does not depend on the path connecting P_1 and P_2 if $\mathbf{A}(x, y, z)$ can be expressed as the gradient of a scalar valued function $\phi(x, y, z)$, i.e. $\mathbf{A}(x, y, z) = \nabla \phi(x, y, z)$. A vector field $\mathbf{A}(x, y, z)$ can be expressed as the gradient of a scalar valued function if the curl of $\mathbf{A}(x, y, z)$ is zero, i.e. $\nabla \times \mathbf{A}(x, y, z) = 0$.

Here we want to point out some issues with respect to differentiation of functions that you should think about.

In calculus you learned about the derivative, and its computation, for scalar valued functions of a scalar variable. So far, we have learned about two generalizations of this idea. We have considered some specific cases of how to differentiate a vector valued function of a scalar variable and a scalar valued function of a vector variable. We recall each case separately. A particular form of a vector valued function of a scalar variable is:

$$\mathbf{A}(t) = A_1(t)\mathbf{i} + A_2(t)\mathbf{j} + A_3(t)\mathbf{k}. \tag{3.45}$$

We use the phrase "particular form" because we are using a particular coordinate system (defined by the unit vectors $\mathbf{i} - \mathbf{j} - \mathbf{k}$) to "represent" the vector, and vectors have a meaning that is independent of the coordinate system in which they are represented.

The derivative of this vector, with respect to t (time for our applications) is given by:

$$\frac{d\mathbf{A}}{dt}(t) = \frac{dA_1}{dt}(t)\mathbf{i} + \frac{dA_2}{dt}(t)\mathbf{j} + \frac{dA_3}{dt}(t)\mathbf{k}. \tag{3.46}$$

So the components of the derivative (in the Cartesian coordinate system) are just the derivative of the components of the original vector. This is easy, but it works because the derivatives of the unit vectors $\mathbf{i}, \mathbf{j}, \mathbf{k}$ are zero (their lengths and directions are constant in time-think of applying the product rule to (3.45) for computing the derivative).

Of course, there are important situations where we represent the vector in a coordinate system defined by unit vectors whose directions vary in time. An important example is the position vector represented in polar coordinates:

$$\mathbf{r}(t) = r(t)\mathbf{r_1}. \tag{3.47}$$

It is important that you know how to differentiate (3.47) twice with respect to time.

Now consider the notion of the derivative of a scalar valued function of a vector variable

$$\phi(\mathbf{r}) = \phi(x, y, z).$$

We have defined the idea of the gradient of such functions (and I hope that you have some appreciation why the idea is useful):

$$\nabla \phi(x, y, z) = \frac{\partial \phi}{\partial x}\mathbf{i} + \frac{\partial \phi}{\partial y}\mathbf{j} + \frac{\partial \phi}{\partial z}\mathbf{k}. \tag{3.48}$$

So the derivative of a vector valued function of a scalar variable is a vector, and the derivative (more precisely, the gradient) of a scalar valued function of a vector is a vector. What is the derivative of a vector valued function of a vector variable? One might think of the curl of a vector valued function of a vector variable as a type of derivative (but this is limited to three, or fewer, dimensions since the cross-product does not generalize to more than three dimensions).

You will be seeing the notion of differentiation many times, and in many forms, throughout your study of mathematics. Fundamentally, in all settings the idea is the same. The derivative of a function (at a point) is a local, linear approximation of the function (near that particular point).

Problem Set 3

1. Let C be a space curve and let the position of any point on the curve be given by the following vector:

$$\mathbf{r} = 3\cos 2t\mathbf{i} + 3\sin 2t\mathbf{j} + (8t - 4)\mathbf{k}.$$

   (a) Find a unit tangent vector $\mathbf{T}$ to the curve (for any point on the curve).
   (b) If $\mathbf{r}$ is the position vector of a particle moving on C at any time t, verify in this case that $\mathbf{v} = v\mathbf{T}$.
   (c) Compute the curvature at any point on the curve.
   (d) Compute the radius of curvature at any point on the curve.
   (e) Compute the unit principal normal $\mathbf{N}$ at any point on the curve.

2. A particle moves so that its position vector is given by:

$$\mathbf{r} = \cos \omega t\mathbf{i} + \sin \omega t\mathbf{j},$$

   where ω is a constant. Prove the following:

   (a) the velocity $\mathbf{v}$ of the particle is perpendicular to $\mathbf{r}$,
   (b) the acceleration $\mathbf{a}$ is directed toward the origin and has magnitude proportional to the distance from the origin,[3]
   (c) $\mathbf{r} \times \mathbf{v}$ is a constant vector.[4]

3. Let $\mathbf{r}(t)$ denote a position vector, and consider the function:

$$T = \frac{1}{2}m\dot{\mathbf{r}} \cdot \dot{\mathbf{r}},$$

   where m is a constant. Compute $\frac{dT}{dt}$.

---

[3]In terms of the vector nature of $\mathbf{a}$, it is important to understand what "directed towards the origin" means, and what "proportional to the distance from the origin" means.

[4]What does "constant vector" mean? Hint: a vector has *length* and *direction*.

4. Let $V(\mathbf{r})$ be a scalar valued function of the position vector $\mathbf{r}(t)$. Compute $\frac{dV}{dt}$, and express it as the dot product of two vectors.

5. Consider the space curve defined by the following position vector:

$$\mathbf{r}(t) = \cos t\,\mathbf{i} + \sin t\,\mathbf{j} + t\,\mathbf{k},$$

and the scalar valued function:

$$V(x, y, z) = \frac{1}{2}\left(x^2 + y^2 + z^2\right).$$

Evaluate the function on the space curve, and then compute its derivative with respect to t.

6. Consider the space curve defined by the following position vector:

$$\mathbf{r}(t) = \cos t\,\mathbf{i} + \sin t\,\mathbf{j} + t\,\mathbf{k}.$$

Compute the length of a piece of the curve from the point $(1, 0, 0)$ to the point $\left(0, 1, \frac{\pi}{2}\right)$.

7. Parametrize the curve in the previous problem in terms of arclength, rather than t.

8. For the example in the previous two exercises show explicitly that:

$$\frac{d\mathbf{r}}{dt} = \frac{d\mathbf{r}}{ds}\frac{ds}{dt}.$$

Chapter 4

Examples of the Computation of Line Integrals and Newton's Axioms

We now consider an example where we compute the line integral of a vector field between two points along three different paths.

Let

$$\mathbf{A}(x, y, z) = x^2\mathbf{i} + y^2\mathbf{j} + z^2\mathbf{k}. \tag{4.1}$$

and consider the points $p_1 = (1, 0, 0)$ and $p_2 = (0, 1, 0)$ shown in Fig. 4.1.

We will compute

$$\int_{p_1}^{p_2} \mathbf{A} \cdot d\mathbf{r}, \tag{4.2}$$

along the following three paths.

1. **Along the path from p_1 to the origin along the x axis, then from the origin to p_2 along the y-axis, see Fig. 4.2.**

   The path is broken into two segments: one piece along the x-axis and the other along the y-axis, see Fig. 4.2. In this case (4.2) becomes:

   $$\int_1^0 x^2 dx + \int_0^1 y^2 dy = 0.$$

2. **Along the straight line in the $x - z$ plane from p_1 to $(0, 0, 1)$, then along the straight line in the $y - z$ plane from $(0, 0, 1)$ to p_2, see Fig. 4.3.**

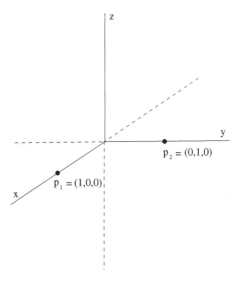

Fig. 4.1. The points p_1 and p_2. The dashed lines represent coordinate axes with negative values.

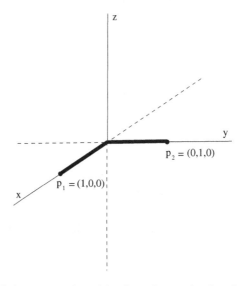

Fig. 4.2. The path from p_1 to the origin along the x-axis, then from the origin to p_2 along the y-axis.

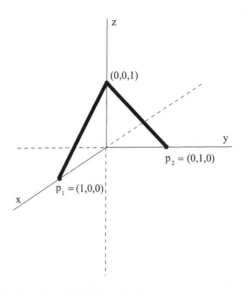

Fig. 4.3.   The path along the straight line in the $x - z$ plane from p_1 to $(0, 0, 1)$, then along the straight line in the $y - z$ plane from $(0, 0, 1)$ to p_2.

This path is also broken into two segments: the line in the $x - z$ plane given by:

$$z = -x + 1, \quad y = 0,$$

and the line in the $y - z$ plane given by:

$$z = -y + 1, \quad x = 0.$$

(In the parametrization of the two paths, make sure you understand the signs on each of the terms describing the path segment.) Then (4.2) becomes:

$$\int_1^0 x^2 dx + (-x + 1)^2(-dx) + \int_0^1 y^2 dy + (-y + 1)^2(-dy) = 0.$$

3. **The path in the $x - y$ plane along $\frac{3}{4}$ of the unit circle from p_1 to p_2, see Fig. 4.4.**

We parametrize the circle by:

$$(x(t), y(t), z(t)) = (\cos t, -\sin t, 0), \quad 0 \le t \le \frac{3\pi}{2}.$$

(Make sure that you understand why this parametrization corresponds to the $\frac{3}{4}$ circle describing the path in the figure.) Then (4.2) becomes:

$$\int_0^{\frac{3\pi}{2}} -\cos^2 t \sin t \, dt - \sin^2 t \cos t \, dt = 0.$$

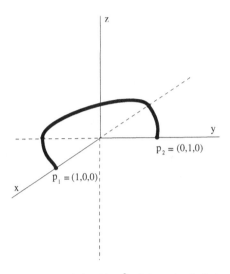

Fig. 4.4. The path from p_1 to p_2 defined by $\frac{3}{4}$ of the unit circle in the $x - y$ plane. The perspective in this figure is awful, just think of the points in the $x - y$ plane that the unit circle must pass through.

It should not have escaped your notice that for all three paths, the value of the line integral was zero. We could verify that the line integral should be independent of the path since the curl of $\mathbf{A}$ is zero. Moreover, we could (probably) determine by inspection that $\mathbf{A} = \nabla\phi(x, y, z)$, where $\phi(x, y, z) = \frac{1}{3}\left(x^3 + y^3 + z^3\right)$. In this case we have:

$$\int_{p_1}^{p_2} \mathbf{A} \cdot d\mathbf{r} = \int_{p_1}^{p_2} \nabla\phi(x, y, z) \cdot d\mathbf{r}$$

$$= \int_{p_1}^{p_2} \left(\frac{\partial\phi}{\partial x}dx + \frac{\partial\phi}{\partial y}dy + \frac{\partial\phi}{\partial z}dz\right)$$

$$= \int_{p_1}^{p_2} d\phi = \phi(1, 0, 0) - \phi(0, 1, 0) = 0. \qquad (4.3)$$

Dynamics–Newton's Axioms

At this point we are finished with our development of purely kinematic ideas and we turn to the study of *dynamics*. A good starting point is to recall Newton's Laws of Motion, which we will take as *axioms*.

Newton's Axioms.

1. Every particle persists in a state of rest or of uniform motion in a straight line (i.e. with constant velocity) unless acted upon by a force.
2. If **F** is the force acting on a particle of mass m which as a consequence is moving with velocity **v**, then

$$\mathbf{F} = \frac{d}{dt}(m\mathbf{v}) = \frac{d\mathbf{p}}{dt}, \tag{4.4}$$

where $\mathbf{p} = m\mathbf{v}$ is called the *momentum*. If m is independent of time t (4.4) becomes:

$$\mathbf{F} = m\frac{d\mathbf{v}}{dt} = m\mathbf{a}, \tag{4.5}$$

where **a** is the acceleration of the particle.
3. If particle 1 acts on particle 2 with a force $\mathbf{F}_{12}$ in a direction along the line joining the two particles, while particle 2 acts on particle 1 with a force $\mathbf{F}_{21}$, then $\mathbf{F}_{21} = -\mathbf{F}_{12}$. In other words, to every *action* there is an equal and opposite reaction.

Example 5. A particle of mass m moves in the $x - y$ plane in such a way that its position vector is given by:

$$\mathbf{r} = a\cos\omega t\mathbf{i} + b\sin\omega t\mathbf{j} \tag{4.6}$$

where a, b, and ω are positive constants, with $a > b$.

1. *Show that the particle moves in an ellipse.*
   From the expression for the position vector we have:

$$x = a\cos\omega t, \quad y = b\sin\omega t,$$

which are just the parametric equations for an ellipse with semi-major axis of length a and semi-minor axis of length b. Since

$$\frac{x^2}{a^2} + \frac{y^2}{b^2} = \cos^2\omega t + \sin^2\omega t = 1,$$

the equation for the ellipse can also be taken as:

$$\frac{x^2}{a^2} + \frac{y^2}{b^2} = 1.$$

2. *Show that the force acting on the particle is always acting towards the origin.*

We will assume that the mass m is constant. Then

$$\mathbf{F} = m\frac{d\mathbf{v}}{dt} = m\frac{d^2\mathbf{r}}{dt^2} = m\frac{d^2}{dt^2}\left(a\cos\omega t\mathbf{i} + b\sin\omega t\mathbf{j}\right),$$

$$= m\left(-\omega^2 a\cos\omega t\mathbf{i} - \omega^2 b\sin\omega t\mathbf{j}\right),$$

$$= -m\omega^2\left(a\cos\omega t\mathbf{i} + b\sin\omega t\mathbf{j}\right) = -m\omega^2\mathbf{r},$$

from which it follows immediately that the force is directed to the origin.

Definitions of Force and Mass. Strictly speaking, force and mass are undefined quantities in Newton's axioms. Intuitively, I expect most of you to have a good idea of what they are. Mass is a measure of the "quantity of matter" in an object. Force is a measure of the "push or pull" on an object. The question of "what is force" and "what is mass" are deep and fundamental. They are questions of current interest in elementary particle physics and quantum field theory (so beyond the scope of this course). It is possible to "define" force and mass through Newton's axioms. Some mechanics books take this approach, but fundamentally, it is not satisfactory (you might think "why not"?).

Units and Dimensions. It is appropriate at this point to say something about dimensions since we will soon be encountering a number of new quantities that are measured with respect to specific types of units.

We saw earlier, that the basic dimensions that we will encounter in this course are length, mass, and time. There will be two systems of units that we will use to describe each: the *centimeter-gram-second (cgs)* system and the *meter-kilogram-second (mks)* system. Units for various quantities are given in Table 4.1.

In terms of units, we can give a definition of force. A *dyne* is the force that will give a 1 gm mass an acceleration of $1\,\mathrm{cm/sec^2}$. A *newton* is the force that will give a 1 kg mass an acceleration of $1\,\mathrm{m/sec^2}$.

Inertial Frames of Reference and Absolute Motion. It needs to be stated that in the course of reasoning from experience that led to Newton's axioms it was always assumed that all measurements or observations were made with respect to a coordinate system or *frame of reference* which was fixed in space, i.e. absolutely at rest. This is the assumption that space or motion is absolute.

Table 4.1.   Units and dimensions.

Physical Quantity	Dimensions	CGS System	MKS System
Length	L	cm	m
Mass	M	gm	kg
Time	T	sec	sec
Velocity	LT^{-1}	cm/sec	m/sec
Acceleration	LT^{-2}	cm/sec^2	m/sec^2
Force	MLT^{-2}	gm cm/sec^2 = dyne	kg m/sec^2 = newton
Momentum, Impulse	MLT^{-1}	gm cm/sec = dyne sec	kg m/sec = nt sec
Energy, Work	ML^2T^{-2}	gm cm^2/sec^2 = dyne cm = erg	kg m^2/sec^2 = nt m = joule
Power	MLT^{-3}	gm cm^2/sec^3 = dyne cm/sec = erg/sec	kg m^2/sec^3 = joule/sec = watt
Volume	L^3	cm^3	m^3
Density	ML^{-3}	gm/cm^3	kg/m^3
Angle	none	radian (rad)	rad
Angular Velocity	T^{-1}	rad/sec	rad/sec
Angular Acceleration	T^{-2}	rad/sec^2	rad/sec^2
Torque	ML^2T^{-2}	gm cm^2/sec^2 = dyne cm	kg m^2/sec^2 = nt m
Angular Momentum	ML^2T^{-1}	gm cm^2/sec	kg m^2/sec
Moment of Inertia	ML^2	gm cm^2	kg m^2
Pressure	$ML^{-1}T^{-2}$	gm/(cm sec^2) = dyne/cm^2	kg/(m sec^2) = nt/m^2

We show that to observers in two different coordinate systems a particle appears to have the same force acting on it *if and only if* the coordinate systems are moving at constant velocity with respect to each other. This is sometimes called *the classical principle of relativity*.

We consider two *observers*, O and O', each located at the origin of a different coordinate system, denoted $x-y-z$ and $x'-y'-z'$, respectively. Each observer observes the motion of a particle P in space. The position vector of P in the $x-y-z$ coordinates system is denoted by $\mathbf{r}$ and the position vector of P in the $x'-y'-z'$ coordinate system is denoted by $\mathbf{r}'$. The vector $\mathbf{R} = \mathbf{r} - \mathbf{r}'$ locates the origin of the $x'-y'-z'$ coordinate system with respect to the $x-y-z$ coordinate system, see Fig. 4.5.

Relative to observers O and O' the forces acting on P according to Newton's laws are given, respectively, by:

$$\mathbf{F} = m\frac{d^2\mathbf{r}}{dt^2}, \quad \mathbf{F}' = m\frac{d^2\mathbf{r}'}{dt^2}. \tag{4.7}$$

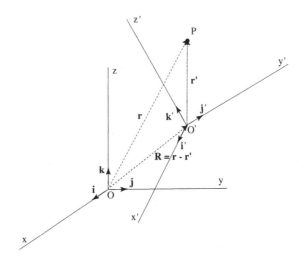

Fig. 4.5. Two distinct coordinate systems in three-dimensional space.

The difference in observed forces is:

$$\mathbf{F} - \mathbf{F}' = m\frac{d^2}{dt^2}(\mathbf{r} - \mathbf{r}') = m\frac{d^2\mathbf{R}}{dt^2}, \tag{4.8}$$

and this will be zero if and only if:

$$\frac{d^2\mathbf{R}}{dt^2} = 0, \quad \text{or} \quad \frac{d\mathbf{R}}{dt} = \text{constant}, \tag{4.9}$$

i.e. the coordinate systems are moving at constant velocity relative to each other. Such coordinate systems are called *inertial coordinate systems*.

Problem Set 4

1. If $\mathbf{A} = (3x^2 - 6yz)\mathbf{i} + (2y + 3xz)\mathbf{j} + (1 - 4xyz^2)\mathbf{k}$ evaluate $\int_C \mathbf{A} \cdot d\mathbf{r}$ from $(0,0,0)$ to $(1,1,1)$ along the following paths C:

   (a) $x = t$, $y = t^2$, $z = t^3$,
   (b) the straight line joining $(0,0,0)$ to $(1,1,1)$.

2. Consider the following scalar valued function of (x, y, z):

$$\phi(x, y, z) = x^3 + zy + xy + \sin xy + \cos\frac{x^2}{z}.$$

   Compute $\frac{\partial \phi}{\partial x}$, $\frac{\partial \phi}{\partial y}$, and $\frac{\partial \phi}{\partial z}$.

3. Suppose

$$\mathbf{A} = (2xy + z^3)\mathbf{i} + (x^2 + 2y)\mathbf{j} + (3xz^2 - 2)\mathbf{k}.$$

Show that the line integral of **A** between two points is independent of the path taken between the two points.

4. Consider the following vector fields:

(a) $\mathbf{A}(x, y, z) = \cos x \sin y \sin z\,\mathbf{i} + \sin x \cos y \sin z\,\mathbf{j} + \sin x \sin y \cos z\,\mathbf{k}$,
(b) $\mathbf{A}(x, y, z) = yz\mathbf{i} + xz\mathbf{j} + xy\mathbf{k}$,
(c) $\mathbf{A}(x, y, z) = z\mathbf{k}$,

Compute the line integral of **A** along any path of your choice connecting $(0, 0, 0)$ to $(1, 1, 1)$.

Chapter 5

Dynamics — Motion of a Particle in One Dimension

"Motion of a Particle in One Dimension" means that the particle is restricted to lie on a curve. If you think hard about this statement you might think that the motion of all particles is one-dimensional since the motion traces out a curve in space. Actually, this is true. However, in order to take advantage of this you must know the curve, i.e. know the motion. In practice, this is usually what we are trying to find out. The situations we have in mind here are when the forces on a particle are always acting so that at each point of a *known curve* they are tangential to that curve. Initially, the "curve" of interest will be a straight line. In this case, the forces act only in the direction of this straight line. We will also consider particles constrained to move in a circle, as well as other plane curves.

We will denote the generic coordinate for the particle by "s". For a particle moving horizontally in the x direction, s would just be x. For a particle falling under the influence of gravity s would be z. For a simple pendulum the particle would be constrained to move in a circle and s would be an angular coordinate, where the angle is measured from some fixed (say vertical) line.

If we assume constant mass, Newton's second axiom of motion becomes:

$$m\frac{d^2 s}{dt^2} = F. \tag{5.1}$$

This is an example of an *ordinary differential equation* (ODE) and all we now need to do is "solve it" to obtain the motion, $s(t)$. However, we are getting a bit ahead of ourselves, because there is the problem of specifying the right-hand side of (5.1), the F. This requires an understanding of the physics of the system under consideration. But first, we want to consider some strictly mathematical issues associated with (5.1).

First some terminology. Newton's equations are *second order* ODEs. This means that the highest order derivatives appearing in the equation are two.

The other point to make is that solutions to second order ODEs require the specification of two constants (which have physical meaning). Why is this? Here is an argument that is not generally found in mechanics textbooks. Suppose $s(t)$ is a solution of (5.1) and we are interested in the solution near $t = t_0$. So we could consider a Taylor expansion near t_0:

$$s(t + t_0) = s(t_0) + \frac{ds}{dt}(t_0)t + \frac{1}{2}\frac{d^2 s}{dt^2}(t_0)t^2 + \frac{1}{6}\frac{d^3 s}{dt^3}(t_0)t^3 + \cdots, \qquad (5.2)$$

and the Taylor coefficients are just the derivatives of $s(t)$ evaluated at t_0. Now in what sense does Eq. (5.1) provide the information to completely specify (5.2)? Here, "completely specify" would mean to completely specify all of the Taylor coefficients. Equation (5.1) provides us with the second derivative of $s(t)$, we can repeatedly differentiate it to obtain all the higher order derivatives. However, it tells us nothing about $s(t_0)$ and $\frac{ds}{dt}(t_0)$. These we must specify separately. If t_0 is the *initial time* we say that we must specify *initial conditions*; the initial position, $s(t_0)$, and the initial velocity, $\dot{s}(t_0)$.

Now for what type of forces can we actually solve (5.1) analytically (as opposed to using a computer)? This depends on the "functional form" of the force, i.e. how it depends on $s, \dot{s}$ and t. We summarize the results in Table 5.1, and then provide some detailed calculations backing up the claims.

Everything was going fine until the last three rows of Table 5.1, and that requires some explanation.

Table 5.1. Examples of mathematical forms of forces for Newton's equations in one dimension.

Force Function	Solvability of Newton's Equations
F = constant	yes
F = F(t)	yes
F = F($\dot{s}$)	yes
F = F(s)	yes
F = F($\dot{s}$, t)	generally no, except for special cases
F= F(s, t)	generally no, except for special cases
F = F(s, $\dot{s}$, t)	generally no, except for special cases

There is an important distinction between linear and nonlinear ODEs. The distinction being that solutions of linear ODEs are fairly simple, while the solutions of nonlinear ODEs *may* be extremely complicated (in ways that can be made mathematically precise).

For these reasons it is important to understand from the outset whether you are dealing with a linear or nonlinear ODE.

First, let us define what we mean by a linear ODE. This means that $F(s, \dot{s}, t)$ is a linear function of s and $\dot{s}$, plus a function solely of t (which could be constant), i.e.

$$F(s, \dot{s}, t) = (a_0 + a_1(t))s + (b_0 + b_1(t))\dot{s} + c_0 + c_1(t), \qquad (5.3)$$

where a_0, b_0, c_0 are constants, and $a_1(t), b_1(t), c_1(t)$ are functions of t.

So are Newton's equations solvable with a force of the form of (5.3)? No. The problem comes from the coefficients on the s and $\dot{s}$ terms. If they are *constant* (i.e. $a_1(t) = b_1(t) = 0$), then Newton's equations can always be solved.

So what is a nonlinear ODE? It is one that is not linear, according to our definition above.

Now let us turn to justifying Table 5.1.

F = constant. Newton's equations are:

$$m\frac{d^2 s}{dt^2} = F = \text{constant}, \quad s(t_0) = s_0, \ \dot{s}(t_0) = v_0,$$

and this is about the easiest differential equation that you could be given to solve. To solve it, you just "integrate twice", as we now show.

$$m \int_{t_0}^{t} \frac{d}{d\tau}\left(\frac{ds}{d\tau}(\tau)\right) d\tau = \int_{t_0}^{t} F d\tau,$$

Performing the integrals gives:

$$\frac{ds}{dt}(t) = v_0 + \frac{F}{m}(t - t_0).$$

Integrating this equation gives:

$$\int_{t_0}^{t} \frac{ds}{d\tau}(\tau)d\tau = v_0 \int_{t_0}^{t} d\tau + \frac{F}{m} \int_{t_0}^{t} (\tau - t_0)d\tau.$$

Performing the integrals gives:

$$s(t) = s_0 + v_0(t - t_0) + \frac{F}{2m}(t - t_0)^2.$$

F = **F**(t). Newton's equations are:

$$m\frac{d^2s}{dt^2} = F(t), \quad s(t_0) = s_0, \quad \dot{s}(t_0) = v_0,$$

and this is about the second easiest differential equation that you could be given to solve. To solve it, you also integrate twice.

$$m\int_{t_0}^{t} \frac{d}{d\tau}\left(\frac{ds}{d\tau}(\tau)\right) d\tau = \int_{t_0}^{t} F(\tau)d\tau,$$

Performing the integrals gives:

$$\frac{ds}{dt}(t) = v_0 + \frac{1}{m}\int_{t_0}^{t} F(\tau)d\tau.$$

Integrating this expression gives:

$$\int_{t_0}^{t} \frac{ds}{d\tau'}(\tau')d\tau' = v_0 \int_{t_0}^{t} d\tau' + \frac{1}{m}\int_{t_0}^{t}\int_{t_0}^{\tau'} F(\tau)d\tau d\tau',$$

or

$$s(t) = s_0 + v_0(t - t_0) + \frac{1}{m}\int_{t_0}^{t}\int_{t_0}^{\tau'} F(\tau)d\tau d\tau'.$$

F = **F**(ṡ). Newton's equations are:

$$m\frac{d^2s}{dt^2} = F\left(\frac{ds}{dt}\right), \quad s(t_0) = s_0, \quad \dot{s}(t_0) = v_0.$$

To solve this equation let

$$u = \frac{ds}{dt}, \tag{5.4}$$

then Newton's equations become:

$$m\frac{du}{dt} = F(u). \tag{5.5}$$

We can solve this equation for $u(t)$ (*provided* we can do the integrals that arise), and then integrate $u(t)$ with respect to t to get $s(t)$.

More precisely, we solve for $u(t)$ by rewriting (5.5) in the following form:

$$m\int_{u(t_0)=v_0}^{u(t)} \frac{du}{F(u)} = \int_{t_0}^{t} d\tau.$$

If this integral can be performed (which will depend on $F(u)$), then we may be able to obtain $u(t) = \frac{ds}{dt}(t)$. We integrate this expression from t_0 to t to obtain $s(t)$.

$\mathbf{F} = \mathbf{F}(\mathbf{s})$. Newton's equations are given by:

$$m\frac{d^2s}{dt^2} = F(s), \quad s(t_0) = s_0, \; \dot{s}(t_0) = v_0. \tag{5.6}$$

Solving this equation requires a certain insight that, fortunately, others have had earlier.

Define the function:

$$V(s) = -\int_c^s F(s')ds', \tag{5.7}$$

where c is any constant. Then Eq. (5.6) becomes:

$$m\frac{d^2s}{dt^2} = -\frac{dV}{ds}(s), \quad s(t_0) = s_0, \; \dot{s}(t_0) = v_0. \tag{5.8}$$

Notice that:

$$\frac{d}{dt}\left(\frac{m}{2}\dot{s}^2 + V(s)\right) = \dot{s}\left(m\ddot{s} + \frac{dV}{ds}(s)\right) = 0. \tag{5.9}$$

Now it is important to interpret this equation correctly. It says that a solution of $m\ddot{s} + \frac{dV}{ds}(s) = 0$ must also satisfy:

$$\frac{m}{2}\dot{s}^2 + V(s) = \text{constant}. \tag{5.10}$$

The constant is determined by the initial conditions. This is a very important expression and later we will see how it is related to "energy".

$\mathbf{F} = \mathbf{F}(\dot{\mathbf{s}}, \mathbf{t})$. If we let

$$\dot{s} = u,$$

then Newton's equations become first order equations for the velocity:

$$m\dot{u} = F(u,t), \quad u(t_0) = v_0.$$

If these equations could be solved then the velocity could be integrated to give the position. Unfortunately, even though they are first order, the general equation cannot be solved explicitly (although many "special cases" are known that can be solved). However, it is true that all *linear first order equations* can be solved, i.e. equations of the form:

$$m\dot{u} = a(t)u + b(t), \quad u(t_0) = v_0, \tag{5.11}$$

where $a(t)$ and $b(t)$ are continuous functions of t. Before demonstrating how this can be solved, let us first simplify this equation by getting rid of the annoying mass term (we can restore it later). We do this by defining:

$$\bar{a}(t) \equiv \frac{a(t)}{m}, \quad \bar{b}(t) \equiv \frac{b(t)}{m}.$$

Then (5.11) becomes:

$$\dot{u} = \bar{a}(t)u + \bar{b}(t), \quad u(t_0) = v_0, \tag{5.12}$$

Now the "trick" to solving this equation is due to Johann Bernoulli. He proposed to write the solution of (5.12) in the form $u(t) = n(t)m(t)$. Substituting this into (5.12) gives:

$$\dot{n}m + \dot{m}n = \bar{a}(t)nm + \bar{b}(t), \quad u(t_0) = n(t_0)m(t_0) = v_0. \tag{5.13}$$

The solution of this equation can be obtained by breaking it into two pieces, and solving each piece separately (why?):

$$\dot{n} = \bar{a}(t)n, \quad \text{solve for } n, \tag{5.14}$$

$$\dot{m} = \frac{\bar{b}(t)}{n}, \quad \text{integrate to get } m, \text{ with } n \text{ substituted in from above.} \tag{5.15}$$

The solution of (5.14) is given by:

$$n(t) = n(t_0)e^{\int_{t_0}^{t} \bar{a}(\tau)d\tau}. \tag{5.16}$$

This is then substituted into (5.15), and integrated, to obtain:

$$m(t) = m(t_0) + \frac{1}{n(t_0)} \int_{t_0}^{t} \bar{b}(\tau)e^{-\int_{t_0}^{\tau} \bar{a}(\tau'')d\tau''} d\tau. \tag{5.17}$$

To obtain $u(t)$, we multiply $n(t)$ and $m(t)$ to obtain:

$$u(t) = n(t)m(t) = n(t_0)m(t_0)e^{\int_{t_0}^{t} \bar{a}(\tau)d\tau}$$
$$+ e^{\int_{t_0}^{t} \bar{a}(\tau)d\tau} \int_{t_0}^{t} \bar{b}(\tau)e^{-\int_{t_0}^{\tau} \bar{a}(\tau'')d\tau''} d\tau, \tag{5.18}$$

and remember that

$$u(t_0) = n(t_0)m(t_0).$$

This expression illustrates the fact that the general solution of (5.12) is the sum of a solution to the homogeneous equation:

$$\dot{u} = \bar{a}(t)u,$$

which is the first term in the solution (5.18), and a particular solution (the second term in the solution (5.18)). A particular solution is just any solution of:

$$\dot{u} = \bar{a}(t)u + \bar{b}(t),$$

where you do not worry about the initial conditions. The initial conditions are then satisfied for the *sum* of the homogeneous plus particular solution

$\mathbf{F} = \mathbf{F}(\mathbf{s}, \mathbf{t})$. Newton's equations are:

$$m\frac{d^2s}{dt^2} = F(s, t), \quad s(t_0) = s_0, \ \dot{s}(t_0) = v_0.$$

Equations of this type cannot generally be solved analytically, *even if they are linear*. However, there is one class of systems of this form which always have a solution: the linear, *constant coefficient* systems, i.e. systems of the form:

$$m\frac{d^2s}{dt^2} = as + b(t),$$

where a is a real number and $b(t)$ is a continuous function of t.

$\mathbf{F} = \mathbf{F}(\mathbf{s}, \dot{\mathbf{s}}, \mathbf{t})$. Newton's equations are:

$$m\frac{d^2s}{dt^2} = F(s, \dot{s}, t), \quad s(t_0) = s_0, \ \dot{s}(t_0) = v_0.$$

Equations of this type cannot generally be solved analytically, *even if they are linear*. However, there is one class of systems of this form which always have a solution: the linear, *constant coefficient* systems, i.e. systems of the form:

$$m\frac{d^2s}{dt^2} = a\dot{s} + bs + c(t),$$

where a and b are real numbers and $c(t)$ is a continuous function of t.

Problem Set 5

1. Consider a projectile that is launched with a velocity of magnitude v_0 at an angle of α with respect to the horizontal. Assume that the only force acting on the projectile is gravity, which is further assumed to be constant, and acting vertically downward (and we may take the vertical coordinate to be z).

   (a) With no loss of generality the motion could be assumed to be in the $y - z$ plane. Explain why.[1]

   (b) Give an argument that the y component of velocity is constant in time.[2]

---

[1] Hint: What does Newton's first law have to say about the situation?
[2] Hint: Again, consider Newton's first law.

2. A particle of mass m moves along a straight line (which, without loss of generality (why?) we may consider to be the x-axis) under the influence of a constant force F, see figure below.

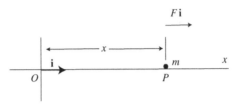

Suppose that the particle starts at $x = 0$ at $t = 0$ with a velocity $v_0\mathbf{i}$ ($v_0 > 0$). Find:

(a) the speed,
(b) the velocity as a function of time,
(c) the distance traveled after time t,
(d) the speed as a function of position (x). (Hint: use the previous two results and eliminate time between them.)

3. An object of mass m is thrown vertically upward from the Earth's surface with an initial velocity $v_0\mathbf{k}$ ($v_0 > 0$). We assume that the only force acting on the object is gravity, see figure below.

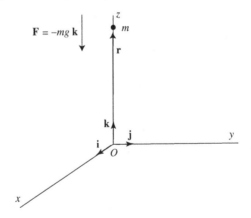

Find:

(a) the position at any time,
(b) the time taken to reach the highest point,
(c) the maximum height reached,
(d) the speed as a function of its distance from the origin.

4. Now we consider a variation of the previous problem. Suppose we drop an object of mass m from a height h at $t = 0$ (from this statement it should be clear that the speed of the object at $t = 0$ is 0). In addition to the force of gravity, suppose there is a force (due to air resistance) directed in the positive z direction that is proportional to the instantaneous speed, i.e. the force has the form $\beta v \mathbf{k}$, where β is a positive scalar, see figure below.

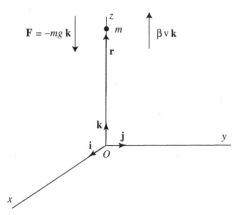

Find:

(a) the speed as a function of time,
(b) the distance traveled as a function of time,
(c) the acceleration at any time $t > 0$.

Show that there is a limiting speed, i.e. as time increases, the speed approaches a limit.

5. Recall the definition of linear ODE given in this chapter:

$$m\frac{d^2 s}{dt^2} = (a_0 + a_1(t))s + (b_0 + b_1(t))\dot{s} + c_0 + c_1(t), \qquad (5.19)$$

where a_0, b_0, c_0 are constants, and $a_1(t), b_1(t), c_1(t)$ are functions of t. First, consider the situation where $c_0 = c_1(t) = 0$, i.e.

$$m\frac{d^2 s}{dt^2} = (a_0 + a_1(t))s + (b_0 + b_1(t))\dot{s}. \qquad (5.20)$$

In this case the linear ODE is said to be homogeneous.

(a) Suppose $s_1(t)$ is a solution of (5.20), and let k_1 denote a constant (real number). Prove that $k_1 s_1(t)$ is also a solution of (5.20).

(b) Suppose $s_1(t)$ and $s_2(t)$ are solutions of (5.20), and let k_1 and k_2 denote constants (real numbers). Prove that $k_1 s_1(t) + k_2 s_2(t)$ is also a solution of (5.20). This is the *superposition principle* for linear homogeneous ODE's.

(c) Do these two results hold for (5.19)?

(d) Now let us consider a physical application of the results in (a) and (b) applied to an ODE of the type (5.20). Consider a particle of mass m moving vertically under the influence of a (constant) gravitational force. Suppose at $t = 0$ the particle is at height 12 (in some units that are not important for this question) with velocity zero. We drop the ball and its position as a function of time is $\tilde{s}(t)$ for $t > 0$.

Now suppose that we consider a different situation. Suppose at $t = 0$ the particle is at height 24 with zero velocity. What is the particle's position as a function of time for $t > 0$?

6. Are the following ordinary differential equations linear or nonlinear?

(a) $m\ddot{s} = -s + \cos t$,

(b) $m\ddot{s} = -s^2 + \cos t$,

(c) $m\ddot{s} = -s \cos t$,

(d) $m\ddot{s} = -t^2 s$,

(e) $m\ddot{s} = -s + s^2$,

7. Solve the following ODE:
$$m\ddot{s} = g, \quad s(0) = s_0, \, \dot{s}(0) = 0,$$
where g is a constant.

8. Solve the following ODE:
$$m\ddot{s} = \sin t, \quad s(0) = s_0, \, \dot{s}(0) = 0.$$

9. Consider the following ODE:
$$m\ddot{s} = s - s^2.$$

Find the function of s and $\dot{s}$ that a solution of this ODE must satisfy.

10. Consider Newton's equations:
$$m\frac{d^2 s}{dt^2} = F(s).$$

Define a new time variable, τ, which is related to the "old" time t by:
$$t = \sqrt{m}\tau.$$

Use the chain rule to show that with respect to the new time the ODE becomes:

$$\frac{d^2 s}{d\tau^2} = F(s),$$

i.e. the constant disappears. This is referred to as *rescaling time*.

11. Consider the following nonlinear ODE:

$$\ddot{s} = s - s^2.$$

Suppose $s_1(t)$ and $s_2(t)$ are solutions, and let k_1 and k_2 denote constants.

(a) Is $k_1 s_1(t)$ a solution?
(b) Is $k_1 s_1(t) + k_2 s_2(t)$ a solution?

Chapter 6

Dynamics — Projectiles, Constrained Motion, Friction

We begin by introducing some new concepts and defining some new terms.

Projectiles. What is a projectile? It is an object that is "projected through space". That is, an object (for us, this will be a particle of constant mass m), at some initial position and time, is given an initial velocity. We are interested in the motion thereafter, which is subject to the applied forces (e.g. gravity, air resistance, etc.). Classically, such problems have been phrased in terms of the flight of cannonballs, bullets, or thrown balls. However, an understanding of the motion of projectiles is fundamental to many current problems in science and engineering. For example, a topic of great recent interest has been concerned with trying to understand how certain rock fragments from the planet Mars made their way to Earth. It turns out that this is a result of the impact of an asteroid with the surface of Mars. The impact dislodged fragments of the planets surface, flinging them out into space. Their subsequent trajectories then came under the influence of the Earth's gravitational field, pulling them onto the surface of the Earth.

Weight and Acceleration Due to Gravity. It has been experimentally determined that, near the Earth's surface, objects fall with a constant vertical acceleration, provided that air resistance is negligible. This acceleration is denoted by **g**, and is referred to as the *acceleration due to gravity* or the *gravitational acceleration*. It is approximately equal to $9.8\,\mathrm{m/sec^2}$.

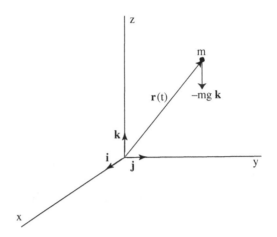

Fig. 6.1. The force of gravity acting on a particle.

Assuming that the Earth's surface is represented by the $x - y$ plane, the force due to gravity acting on a particle of mass m is given by:

$$\mathbf{W} = -mg\mathbf{k}.$$

This force, which is referred to as *the weight of the particle*, has magnitude $W = mg$, see Fig. 6.1.

We remark that this assumption of a *flat Earth* is only approximately correct. Moreover, the force of gravity varies slightly over the surface of the Earth. You will learn more about this when you learn about *Newton's Law of Gravitation*.

Constrained Motion. In some cases a particle is forced to move along a curve or surface. This curve or surface is referred to as a *constraint*, and the resulting motion is called *constrained motion*.

The particle exerts a force on the constraint, and by Newton's third law the constraint exerts a force on the particle. This force is called the *reaction force*, and is described by giving its components normal to the motion, denoted $\mathbf{N}$, and parallel to the motion, denoted $\mathbf{f}$.

Friction. In the constrained motion of a particle a common force parallel to the motion of the particle is *friction*, which typically acts in the direction opposite to that of the motion (i.e. it acts in such a way as to slow the motion of the particle). Experimentally, it is found that the magnitude of the frictional force is proportional to the magnitude of the normal force, i.e.

$$f = \mu N,$$

where the proportionality constant, μ is referred to as the *coefficient of friction*. Typically, its value depends on both the material of the particle and the constraint.

Example 6. Consider a projectile of constant mass m that is launched with initial speed v_0 at an angle α with the horizontal. We will compute:

1. the position vector at any time,
2. the time to reach the highest point,
3. the maximum height reached,
4. the time of flight back to earth,
5. the range.

For an illustration of the geometry see Fig. 6.2.

We will denote the position vector of the projectile at any time t by $\mathbf{r}$, and the velocity at any time t by $\mathbf{v}$. Of course, we begin by writing down Newton's equations:

$$m\frac{d^2\mathbf{r}}{dt^2} = -mg\mathbf{k},$$

which, after dividing both sides by the mass, become:

$$\frac{d^2\mathbf{r}}{dt^2} = -g\mathbf{k}, \quad \text{or} \quad \frac{d\mathbf{v}}{dt} = -g\mathbf{k}.$$

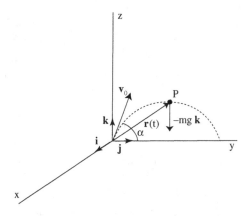

Fig. 6.2. Geometry of the motion of a projectile under the influence of gravity.

The equation for **v** is easily solved to give:

$$\mathbf{v} = -gt\mathbf{k} + \mathbf{c}_1. \tag{6.1}$$

To evaluate the constant vector $\mathbf{c}_1$ we need to require our solution to satisfy the initial conditions. We assume that our initial velocity is in the $y - z$ plane (why is it sufficient to assume this?) so that the initial conditions take the form:

$$\mathbf{v}_0 = v_0 \cos \alpha \mathbf{j} + v_0 \sin \alpha \mathbf{k}. \tag{6.2}$$

Evaluating (6.1) at $t = 0$ and equating the result to $\mathbf{v}_0$ gives $\mathbf{c}_1 = \mathbf{v}_0$, and therefore we have:

$$\mathbf{v} = v_0 \cos \alpha \mathbf{j} + (v_0 \sin \alpha - gt)\mathbf{k}. \tag{6.3}$$

Now we can solve for the **position vector**. Writing $\mathbf{v} = \frac{d\mathbf{r}}{dt}$, substituting this into (6.3), and integrating from 0 to t (using the initial condition $\mathbf{r}(0) = 0$), gives:

$$\mathbf{r} = (v_0 \cos \alpha)t\mathbf{j} + \left((v_0 \sin \alpha)t - \frac{g}{2}t^2 \right) \mathbf{k}. \tag{6.4}$$

or, in components,

$$x = 0, \quad y = (v_0 \cos \alpha)t, \quad z = (v_0 \sin \alpha)t - \frac{g}{2}t^2. \tag{6.5}$$

So we see that if the projectile has initial velocity in the $y - z$ plane, it remains in the $y - z$ plane. Should we have been able to determine this without solving the equations of motion?

Next we compute **the time to reach the highest point**. What characterizes the highest point? The projectile goes up, then it comes down. In order to come down, the z component of velocity must change from positive to negative. Therefore, at the highest point, the z component of velocity vanishes, i.e.

$$v_0 \sin \alpha - gt = 0, \quad \text{and therefore} \quad t = \frac{v_0 \sin \alpha}{g}.$$

Now we compute **the maximum height reached**. What characterizes the maximum height? This is the same as the previous question. The height is a maximum when the z component of velocity vanishes. We computed the time for this to occur in the previous question. We then substitute this time into the expression for the z component of position given n (6.5) to obtain:

$$\text{maximum height reached} = (v_0 \sin \alpha) \left(\frac{v_0 \sin \alpha}{g} \right) - \frac{g}{2} \left(\frac{v_0 \sin \alpha}{g} \right)^2$$

$$= \frac{v_0^2 \sin^2 \alpha}{2g}.$$

Next we compute **the time of flight back to Earth**. What characterizes this time? This is the time for the z value to return to $z = 0$. To obtain this we take the expression for the z component of the motion as a function of time given in (6.5) and equate it to zero, solving for t:

$$(v_0 \sin \alpha)t - \frac{g}{2}t^2 = t \left(v_0 \sin \alpha - \frac{g}{2}t \right) = 0,$$

which gives:

$$t = \frac{2v_0 \sin \alpha}{g}.$$

$t = 0$ is also a solution. Why do we discount it?). Note that this is twice the value that it takes to reach the maximum height. Would you expect that?

Finally, we compute **the range**. This is the maximum distance that the projectile travels horizontally (i.e. in the y direction). To compute this we substitute the time that it takes to return back to Earth into the expression for the y component given in (6.5):

$$\text{Range} = (v_0 \cos \alpha) \left(\frac{2v_0 \sin \alpha}{g} \right) = \frac{2v_0^2 \sin \alpha \cos \alpha}{g} = \frac{v_0^2 \sin 2\alpha}{g}.$$

Example 7. A particle P of constant mass m slides without rolling down a frictionless inclined plane of angle α, see figure below.

If the particle starts from rest at the top of the incline (at point A) find:

1. the acceleration,
2. the velocity,
3. the distance traveled after time t.

Since there is no friction the only forces acting on the particle are the weight $\mathbf{W} = -mg\mathbf{k}$ and the reaction force of the incline which is given by the normal force $\mathbf{N}$.

Let $\mathbf{e}_1$ and $\mathbf{e}_2$ be unit vectors parallel and perpendicular to the incline, respectively. We let s denote the magnitude of the displacement of the

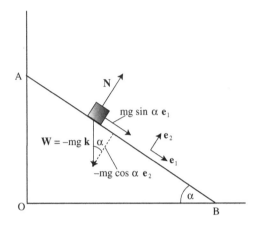

Fig. 6.3. Geometry of the motion of a particle under the influence of gravity down an inclined plane.

particle measured from the top of the incline (point A). Now we can write down Newton's second law for this system:

$$m\frac{d^2}{dt^2}(s\mathbf{e}_1) = \mathbf{W} + \mathbf{N} = mg\sin\alpha\mathbf{e}_1,$$

or

$$\frac{d^2s}{dt^2} = g\sin\alpha,$$

which gives the **acceleration**.

Next we compute the **speed**. Since $\frac{ds}{dt} = v$ we have:

$$\frac{dv}{dt} = g\sin\alpha,$$

or

$$v = (g\sin\alpha)t + c_1$$

Using the initial condition $v = 0$ at $t = 0$, we obtain $c_1 = 0$. Hence, the speed at any time t is given by:

$$v = (g\sin\alpha)t.$$

The **velocity** at any time t is given by:

$$v\mathbf{e}_1 = (g\sin\alpha)t\mathbf{e}_1.$$

Finally, we compute **the distance traveled after time** t. Since $v = \frac{ds}{dt}$ we have:

$$\frac{ds}{dt} = (g\sin\alpha)t \quad \text{or} \quad s = \frac{1}{2}(g\sin\alpha)t^2 + c_2.$$

Using the initial condition $s = 0$ at $t = 0$ we find that $c_2 = 0$, and therefore

$$s = \frac{1}{2}(g \sin \alpha)t^2.$$

We continue these topics with a further example.

Example 8. A particle, denoted by P, sits at the top (point A in the figure) of a frictionless fixed sphere of radius b, see Fig. 6.4.

The particle is displaced slightly so that it slides (without rolling) down the sphere.

Compute:

1. the position of the particle as it leaves the sphere,
2. the speed of the particle at the instant it leaves the sphere.

The particle slides along a circle which, as indicated in the figure, we have taken to lie in the $x - y$ plane. The forces acting on the particle are its

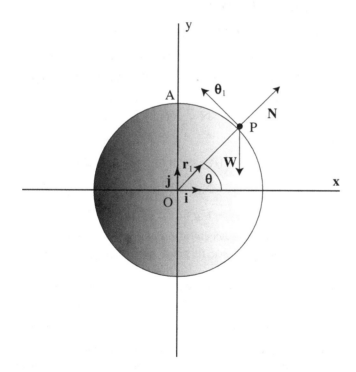

Fig. 6.4. Particle moving on a sphere under the influence of gravity.

weight, $\mathbf{W} = -mg\mathbf{j}$, and the reaction force $\mathbf{N}$ of the sphere on the particle normal to the sphere.

We let the position of the particle on the circle be measured by the angle θ, and we let $\mathbf{r}_1$ and $\boldsymbol{\theta}_1$ denote unit vectors in the radial direction and the direction of increasing θ, respectively (see Fig. 6.4).

For this problem it is most natural to use a (moving) coordinate system tangential and normal to the sphere since the question of interest is how far the particle has moved along the sphere when it leaves the sphere.

Therefore, we want to express Newton's equations in this coordinate system. We begin by resolving the forces into components along these directions. First, as a preliminary calculation, note from Fig. 6.4, that we have:

$$\mathbf{j} \cdot \mathbf{r}_1 = |\mathbf{j}| \, |\mathbf{r}_1| \sin \theta = \sin \theta,$$

$$\mathbf{j} \cdot \boldsymbol{\theta}_1 = |\mathbf{j}| \, |\boldsymbol{\theta}_1| \cos \theta = \cos \theta,$$

We use these relations to resolve the force due to gravity on the particle in the $\mathbf{r}_1$ and $\boldsymbol{\theta}_1$ directions. This is given by:

$$\mathbf{W} = (\mathbf{W} \cdot \mathbf{r}_1)\mathbf{r}_1 + (\mathbf{W} \cdot \boldsymbol{\theta}_1)\boldsymbol{\theta}_1,$$

$$= (-mg\mathbf{j} \cdot \mathbf{r}_1)\mathbf{r}_1 + (-mg\mathbf{j} \cdot \boldsymbol{\theta}_1)\boldsymbol{\theta}_1,$$

$$= -mg \sin \theta \mathbf{r}_1 - mg \cos \theta \boldsymbol{\theta}_1.$$

The normal reaction force clearly acts in radial direction (since this is normal to the sphere):

$$\mathbf{N} = N\mathbf{r}_1.$$

Hence, Newton's laws become:

$$\mathbf{F} = m\mathbf{a} = m\left((\ddot{r} - r\dot{\theta}^2)\mathbf{r}_1 + (r\ddot{\theta} + 2\dot{r}\dot{\theta})\boldsymbol{\theta}_1 \right),$$

$$= \mathbf{W} + \mathbf{N},$$

$$= (N - mg \sin \theta)\mathbf{r}_1 - mg \cos \theta \boldsymbol{\theta}_1,$$

where the expression for the acceleration is given by the second derivative of the position vector *expressed in polar coordinates* which was discussed earlier.

Alternately, we can write the equations separately in components as follows:

$$m(\ddot{r} - r\dot{\theta}^2) = N - mg \sin \theta, \quad m(r\ddot{\theta} + 2\dot{r}\dot{\theta}) = -mg \cos \theta. \tag{6.6}$$

But in this problem, the particle is constrained to move along the sphere (until it leaves, but we will not be concerned with the motion after this happens), so we have $r = b$, and therefore $\dot{r} = \ddot{r} = 0$. Therefore, the equations of motion simplify to:

$$-mb\dot{\theta}^2 = N - mg\sin\theta, \quad b\ddot{\theta} = -g\cos\theta. \tag{6.7}$$

Now that we have written down the equations of motion we have to think what we have been asked to compute. The first thing we want to know is **at what position (i.e. value of θ) will the particle leave the sphere?** We must ask ourselves what would characterize this event (in terms of the applied forces)? The particle leaves the sphere when the normal reaction force is zero (think about this). Therefore, we would like to find an expression for the normal force as a function of θ, then equate the expression to zero and solve for θ.

How do we find an expression for the normal force? We have to go to the equations of motion (6.7), and be a bit clever.

If we multiply the second equation in (6.7) by $\dot{\theta}$, we see that the result can be written as:

$$b\frac{d}{dt}\left(\frac{\dot{\theta}^2}{2}\right) = -g\frac{d}{dt}(\sin\theta).$$

Integrating this equation gives:

$$b\frac{\dot{\theta}^2}{2} = -g\sin\theta + c.$$

Now at the top of the sphere, $\theta = \frac{\pi}{2}$, we have $\dot{\theta} = 0$. Therefore $c = g$, and we have:

$$b\dot{\theta}^2 = 2g(1 - \sin\theta). \tag{6.8}$$

Substituting this expression into the first equation of (6.7) gives:

$$N = mg(3\sin\theta - 2),$$

which is our sought after expression for the normal force as a function of θ. Equating this expression to zero gives:

$$\sin\theta = \frac{2}{3}, \quad \text{or} \quad \theta = \sin^{-1}\frac{2}{3}. \tag{6.9}$$

Finally, we want to find **the speed of the particle at the moment it leaves the sphere**. The angular speed of the particle while it is on the sphere is given by $|\dot{\theta}|$.

If we substitute (6.9) into (6.8) we obtain:

$$\dot{\theta}^2 = \frac{2g}{3b}.$$

So the angular speed of the particle when it leaves the sphere is given by:

$$|\dot{\theta}| = \sqrt{\frac{2g}{3b}}.$$

Therefore the speed of the particle when it leaves the sphere is given by:

$$|b\dot{\theta}| = \sqrt{\frac{2}{3}gb}.$$

Finally, we end this chapter with some general remarks. In particular, we address the issue of:

"How do you solve mechanics problems?"

Initially, this might seem to be a curious statement since the point of the lectures is to provide you with the tools for solving mechanics problems. They do provide you with the tools. However, in the first year many students find the problems in mechanics more difficult than those in, say, calculus. There is probably some truth to this, so we should look at this in more detail.

First, in calculus many of the problems that you are set involve performing a specific computation that should be clear from the statement of the problem. For example, you might be given a scalar valued function of two variables and asked to compute its (first) partial derivatives. Hopefully, you understand that the solution of that problem is a straightforward calculation. Another example involves integration of a function of one variable. You may be given a function and asked to integrate it between two limits. What you are asked to do should be straightforward. However, some insight (coming from experience) is required in order to know what technique to employ (e.g. partial fractions, integration by parts, etc.) to successfully integrate the given function. Knowing how to start with a mechanics problem is essential to answering the questions related to the problem that you are asked. Let us consider this in the context of the first two examples in this lecture.

The first example concerns the motion of a projectile. You are given a one sentence description of the problem, and then asked to determine five quantities associated with the motion of the projectile. The natural thing to do is to focus on the five quantities that you are asked to determine.

But this is not where you start. Instead, you should first focus on writing down Newton's equations, i.e. the second order differential equation that is a consequence of Newton's second law of motion. This involves determining the forces acting on the particle (at the moment we are concerning with one particle of constant mass m). Once this is done, you can write down Newton's equations as a vector equation (the mass multiplied by the second derivative of the position vector is equal to the vector sum of all of the forces acting on the particle). Now the motion tends to be described in terms of a particular coordinate system (e.g. how high does the particle go? how far does it go? where is it along the inclined plane?). In order to do this you will need to choose a set of coordinates that are appropriate to the problem in which to represent Newton's equations, i.e. represent the vector quantities in Newton's equations in components with respect to the chosen coordinates. Once you have carried out these steps (determine the forces acting on the particle, write down Newton's equations, choose an appropriate set of coordinates in which to represent Newton's equations) then you will be ready to answer the questions concerning the motion of the particle that you are asked. However, note that when determining Newton's equations (i.e. the nature of the vector sum of forces acting on the particle) you may have to use Newton's first and third laws of motion. See how this arises in the examples in this chapter.

Problem Set 6

Key point: The are several points to take particular care with in the problems below.

- Do the distances and times that you compute make "physical sense" in the context of the stated problem? For example, are particles moving through solid obstacles? Are time intervals real and positive?
- In some cases you will have to solve quadratic equations, which have two solutions. Make sure you determine if both solutions, one solution, or no solutions make "physical sense" in the context of the stated problem.
- Distances are measured with respect to a chosen coordinate system. Make sure you are taking the appropriate signs for a distance with respect to the chosen coordinate system.

1. A particle P of constant mass m slides without rolling down an inclined plane of angle α that has a constant coefficient of friction μ, see figure below.

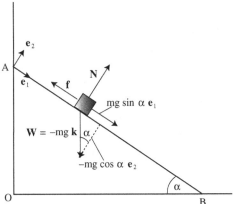

If the particle starts from rest at the top of the incline (at point A) find:

(a) the acceleration,
(b) the velocity,
(c) the distance traveled after time t.

2. Refer to figure below. An inclined plane makes an angle α with the horizontal.

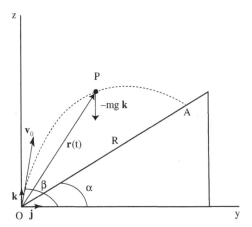

A projectile is launched from the bottom of the plane (point O) with speed v_0 in a direction making an angle β with the horizontal. Prove that the range R up the incline is given by:

$$R = \frac{2v_0^2 \sin(\beta - \alpha) \cos \beta}{g \cos^2 \alpha}.$$

3. An object slides on a surface along the horizontal straight line OA, see figure below.

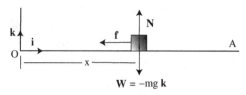

$$W = -mg\,\mathbf{k}$$

We assume that $x = 0$ and $v = v_0$ at $t = 0$. Suppose that the object comes to rest after traveling a distance x_0. Show that the coefficient of friction is:

$$\frac{v_0^2}{2gx_0}.$$

4. This question is concerned with the first example, the projectile problem, from Chapter 6. The expression for the vertical displacement of the particle as a function of time is given by:

$$z(t) = (v_0 \sin \alpha)\,t - \frac{g}{2}t^2.$$

Clearly, for t sufficiently large, $z(t)$ can be negative. Does this pose any problems?

5. The set-up for this question is the same as the first example, the projectile problem, from Chapter 6, *except* at a distance d from the launch point of the projectile, the horizontal boundary suddenly "drops" to a distance H below zero, see figure below (think of launching the projectile from a cliff).

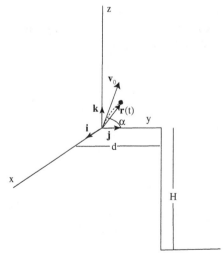

Compute the time that it takes the projectile to reach the vertical distance H below the launch point.

6. The set-up for this question is the same as the first example, the projectile problem, from Chapter 6, *except* at a distance d from the launch point of the projectile, a wall is placed of height H, see figure below (the wall is parallel to the $x - z$ plane).

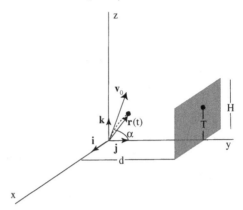

(a) Compute the time that it takes the projectile to hit the wall.

(b) Show that a necessary condition for a particle to hit the wall at a height $T > 0$ is:

$$d < \frac{2v_0^2 \sin \alpha \cos \alpha}{g}.$$

(c) Now suppose $\alpha = 45^0$, $v_0 = 100\,\text{m/s}$, and $g = 9.8\,\text{m/s}^2$. With these parameters fixed, what value of d must we take in order to hit the wall at a height of $10\,\text{m}$?

Chapter 7

Dynamics — Work and Power

We now introduce the ideas of *work* and *power*. The notion of work can be viewed as the "bridge" between Newton's second law, and energy (which we have yet to define and discuss). The term work was first used in 1826 by the French mathematician/engineer Gaspard–Gustave Coriolis (note that this was, roughly, 150 years after Newton).

Work. Suppose a force $\mathbf{F}$ acting on a particle gives it a displacement $d\mathbf{r}$. Then the work done by the force on the particle is defined as:

$$dW = \mathbf{F} \cdot d\mathbf{r}. \tag{7.1}$$

It should be clear that work is a scalar.) The total work done by a force field (or, vector field) $\mathbf{F}$ in moving a particle from point P_1 to point P_2 along a path C is given by the line integral:

$$W = \int_C \mathbf{F} \cdot d\mathbf{r} = \int_{P_1}^{P_2} \mathbf{F} \cdot d\mathbf{r} = \int_{\mathbf{r}_1}^{\mathbf{r}_2} \mathbf{F} \cdot d\mathbf{r}, \tag{7.2}$$

where $\mathbf{r}_1$ is the position vector of P_1 and $\mathbf{r}_2$ is the position vector of P_2, see Fig. 7.1.

It is worth considering $\mathbf{F}$ more carefully in the expression for work. In our discussion of Newton's second law, $\mathbf{F} = m\mathbf{a}$, $\mathbf{F}$ was the vector sum of all forces acting on the particle of mass m, i.e. the "net force acting on m". This will "almost always" be the case in our expressions for work, and it is definitely the case when Newton's second law is used in certain manipulations involving the expression for work, e.g. in Theorem 1. However, there

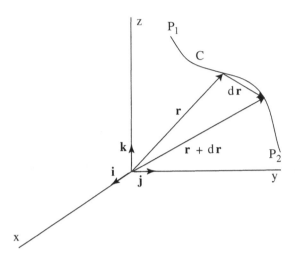

Fig. 7.1. Geometry associated with a force acting on a particle in three-dimensional space.

may be circumstances when one is interested in the work done on a particle *by a specific force* in the vector sum of all forces. This should be clear from the context. Unless explicitly stated, the **F** in the expression for work is the vector sum of all forces acting on the particle of constant mass m.

The words of Sommerfeld[1] are particularly insightful when considering the concept of work.

> Work does not equal 'force times distance' as often stated, but 'component of force along path times path length' or 'force times component of path length along force'.

Now consider carefully Sommerfeld's statement in the context of the following example.

Example 9. Find the work done in moving an object along a vector $\mathbf{r} = 3\mathbf{i} + 2\mathbf{j} - 5\mathbf{k}$ if the applied force is $\mathbf{F} = 2\mathbf{i} - \mathbf{j} - \mathbf{k}$, see Fig. 7.2 (following Sommerfeld, this means "along a distance of constant length of the vector in the constant direction of the vector").

---

[1] Arnold Sommerfeld, MO Stern, and RB Lindsay. *Mechanics*. Lectures on Theoretical Physics, Volume i, 1954.

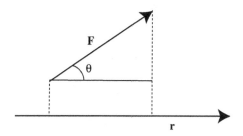

Fig. 7.2. An example of a force acting on a particle.

Solution:

Work done = magnitude of force in direction of motion × distance moved,

$$= (F\cos\theta)(r) = \mathbf{F} \cdot \mathbf{r},$$

$$= (2\mathbf{i} - \mathbf{j} - \mathbf{k}) \cdot (3\mathbf{i} + 2\mathbf{j} - 5\mathbf{k}),$$

$$= 6 - 2 + 5 = 9.$$

We consider another example.

Example 10. Find the work done in moving a particle once around a circle C in the $x - y$ plane, if the circle has center at the origin and radius 3, and if the force field is given by:

$$\mathbf{F} = (2x - y + z)\mathbf{i} + (x + y - z^2)\mathbf{j} + (3x - 2y + 4z)\mathbf{k},$$

see Fig. 7.3.

Solution: In the plane $z = 0$, $\mathbf{F} = (2x - y)\mathbf{i} + (x + y)\mathbf{j} + (3x - 2y)\mathbf{k}$ and $d\mathbf{r} = dx\mathbf{i} + dy\mathbf{j}$. Then the work done is:

$$\int_C \mathbf{F} \cdot d\mathbf{r} = \int_C ((2x - y)\mathbf{i} + (x + y)\mathbf{j} + (3x - 2y)\mathbf{k}) \cdot (dx\mathbf{i} + dy\mathbf{j}),$$

$$= \int_C (2x - y)dx + (x + y)dy.$$

Choose the parametric equations of the circle as $x = 3\cos t$, $y = 3\sin t$, where t varies from 0 to 2π. Then the line integral becomes:

$$\int_{t=0}^{2\pi} (2(3\cos t) - 3\sin t)(-3\sin t)\, dt + (3\cos t + 3\sin t)(3\cos t)dt,$$

$$= \int_0^{2\pi} (9 - 9\sin t \cos t)dt = \left(9t - \frac{9}{2}\sin^2 t\right)\Big|_0^{2\pi} = 18\pi.$$

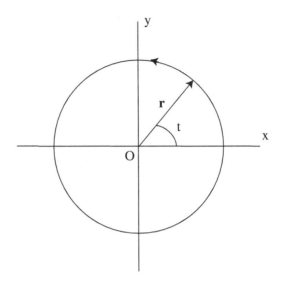

Fig. 7.3. A force moving a particle around a circle.

In traversing C we have chosen the counterclockwise direction indicated in Fig. 7.3. We refer to this as the *positive direction* and say that C *has been traversed in the positive sense.* If we had traversed C in the clockwise (negative) direction the value of the integral would have been -18π.

Power. The time rate of doing work on a particle is called the *instantaneous power*, or, briefly, the *power* applied to the particle. Denoting the instantaneous power by $\mathcal{P}$, we have:

$$\mathcal{P} = \frac{dW}{dt}.$$

If $\mathbf{F}$ denotes the force acting on a particle and $\mathbf{v}$ is the velocity of the particle then we also have:

$$\mathcal{P} = \mathbf{F} \cdot \mathbf{v}.$$

Example 11. Suppose a particle of mass m moves under the influence of a force field along a space curve whose position vector is given by:

$$\mathbf{r} = (2t^3 + t)\mathbf{i} + (3t^4 - t^2 + 8)\mathbf{j} - 12t^2\mathbf{k}.$$

Find the instantaneous power applied to the particle by the force field.

Solution: We need to compute the force and the velocity, and then take their dot product. The velocity of the particle is given by:

$$\mathbf{v} = (6t^2 + 1)\mathbf{i} + (12t^3 - 2t)\mathbf{j} - 24t\mathbf{k}.$$

The acceleration is given by:

$$\mathbf{a} = 12t\mathbf{i} + (36t^2 - 2)\mathbf{j} - 24\mathbf{k}.$$

Therefore the force is given by:

$$\mathbf{F} = m\mathbf{a} = m\left(12t\mathbf{i} + (36t^2 - 2)\mathbf{j} - 24\mathbf{k}\right).$$

Now we have enough information to compute the instantaneous power:

$$\mathbf{F} \cdot \mathbf{v} = m\left(12t\mathbf{i} + (36t^2 - 2)\mathbf{j} - 24\mathbf{k}\right) \cdot \left((6t^2 + 1)\mathbf{i} + (12t^3 - 2t)\mathbf{j} - 24t\mathbf{k}\right),$$
$$= m\left(72t^3 + 12t + 432t^5 - 72t^3 - 24t^3 + 4t + 576t\right)$$
$$= m\left(432t^5 - 24t^3 + 592t\right).$$

Dynamics–Kinetic Energy

We now show that the idea of work leads naturally to the notion of the idea of the *kinetic energy of a particle*. If the force is conservative, we show that this enables us to introduce the idea of *potential energy*. In this way work can be characterized either through kinetic energy, or through the potential energy.

Kinetic Energy. Refer back to Fig. 7.4 which we reproduce again here for ease of reference.

We assume that the mass m of the particle is constant, and that at times t_1 and t_2, it is located at points P_1 and P_2, respectively, and moving with velocities $\mathbf{v}_1 = \frac{d\mathbf{r}}{dt}(t_1)$ and $\mathbf{v}_2 = \frac{d\mathbf{r}}{dt}(t_2)$ at points P_1 and P_2, respectively. Then we can prove the following theorem.

Theorem 1. *The total work done by the net forces* $\mathbf{F}$ *in moving the particle along the curve C from P_1 to P_2 is given by:*

$$W = \int_C \mathbf{F} \cdot d\mathbf{r} = \frac{1}{2}m(v_2^2 - v_1^2). \tag{7.3}$$

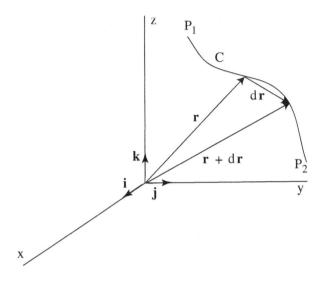

Fig. 7.4.   Geometry of a force acting on a particle in three-dimensional space.

Proof.

$$\text{Work done} = \int_{t_1}^{t_2} \mathbf{F} \cdot \frac{d\mathbf{r}}{dt} dt = \int_{t_1}^{t_2} \mathbf{F} \cdot \mathbf{v} dt,$$

$$= \int_{t_1}^{t_2} m \frac{d\mathbf{v}}{dt} \cdot \mathbf{v} dt = m \int_{t_1}^{t_2} \mathbf{v} \cdot d\mathbf{v},$$

Note that Newton's second law is used here.

$$= \frac{1}{2} m \int_{t_1}^{t_2} d(\mathbf{v} \cdot \mathbf{v}) = \frac{1}{2} m v^2 \Big|_{t_1}^{t_2} = \frac{1}{2} m v_2^2 - \frac{1}{2} m v_1^2,$$

which completes the proof.[2]   □

If we call the quantity:

$$T = \frac{1}{2} m v^2,$$

the *kinetic energy*, this theorem states that:

The total work done by the net forces $\mathbf{F}$ in moving the particle of constant mass m from P_1 to P_2 along C = the kinetic energy at P_2 — the kinetic energy at P_1.

---

[2]It is important to be able to justify all steps in this calculation to yourself.

or, symbolically,

$$W = T_2 - T_1,$$

where $T_1 = \frac{1}{2}mv_1^2$ and $T_2 = \frac{1}{2}mv_2^2$.

It should be clear (but you should convince yourself of this) that C is a solution of Newton's equations.

Now we consider the important case when the force is *conservative*.

Conservative Force Fields. Suppose there exists a scalar function V such that $\mathbf{F} = -\nabla V$. Then we can prove the following:

Theorem 2. *The total work done by the force* $\mathbf{F} = -\nabla V$. *in moving the particle along C from P_1 to P_2 is:*

$$W = \int_{P_1}^{P_2} \mathbf{F} \cdot d\mathbf{r} = V(P_1) - V(P_2).$$

Proof. First, note that:

$$dV(x, y, z) = \frac{\partial V}{\partial x}(x, y, z)dx + \frac{\partial V}{\partial y}(x, y, z)dy + \frac{\partial V}{\partial z}(x, y, z)dz = \nabla V \cdot d\mathbf{r}.$$

Then the proof follows from a direct calculation using the specific form of $\mathbf{F} = -\nabla V$:

$$W = \int_{P_1}^{P_2} \mathbf{F} \cdot d\mathbf{r} = \int_{P_1}^{P_2} -\nabla V \cdot d\mathbf{r} = \int_{P_1}^{P_2} -dV = -V \Big|_{P_1}^{P_2} = V(P_1) - V(P_2),$$

which completes the proof. $\square$

Now we state (without proof) two theorems that give computable conditions for determining if a force field is conservative.

Theorem 3. *A force field $\mathbf{F}$ is conservative if and only if there exists a continuously differentiable scalar field V such that $\mathbf{F} = -\nabla V$ or, equivalently, if and only if:*

$$\nabla \times \mathbf{F} = \operatorname{curl} \mathbf{F} = 0 \quad \text{identically.}$$

Theorem 4. *A continuously differentiable force field $\mathbf{F}$ is conservative if and only if for any closed nonintersecting curve C (i.e. simple closed curve) we have:*

$$\oint_C \mathbf{F} \cdot d\mathbf{r} = 0,$$

i.e. the total work done in moving a particle around any simple closed curve is zero.

Potential Energy or Potential. The scalar function V, such that $\mathbf{F} = -\nabla V$ is called the *potential energy* (or also the *scalar potential* or just *potential*) of the particle in the conservative force field $\mathbf{F}$.

It should be noted that if you add an arbitrary constant to the potential, the associated force does not change. We can express the potential as:

$$V = - \int_{r_0}^{r} \mathbf{F} \cdot d\mathbf{r}',$$

where r_0 is chosen arbitrarily. The point r_0 is sometimes called the *reference point*. Even though it is arbitrary, a wise choice can simplify a problem. We will see examples of this later on.

Problem Set 7

1. Suppose $\mathbf{F}$ is the net force acting on a particle of constant mass m, and suppose that $\mathbf{F}$ gives the particle a displacement $d\mathbf{r}$. Then, we defined the work done by $\mathbf{F}$ in moving the particle through a displacement $d\mathbf{r}$ as $dW = \mathbf{F} \cdot d\mathbf{r}$. Recall that Newton's second law is $\mathbf{F} = m\frac{d^2\mathbf{r}}{dt^2}$, i.e. $\mathbf{F}$ is proportional to the second derivative of the displacement $\mathbf{r}$ of the particle. Then shouldn't $\mathbf{F}$ be proportional to the displacement $\mathbf{r}$?

2. Let $\mathbf{r}$ denote the position vector of a particle of constant mass m and let $\mathbf{v} = \frac{d\mathbf{r}}{dt}$ denote the velocity of the particle. Let $\mathbf{A}$ denote a vector such that $\mathbf{F} = \mathbf{v} \times \mathbf{A}$ is the net force acting on the particle. Show that this force does no work on the particle.

3. Explain the following statements:

   (a) Work is equal to the transference of kinetic energy.
   (b) There can be no work without motion.

4. A particle of mass m moves in the $x - y$ plane so that its position vector is:

$$\mathbf{r} = a\cos\omega t\mathbf{i} + b\sin\omega t\mathbf{j},$$

where a, b, and ω are positive constants with $a > b$, see the following figure.

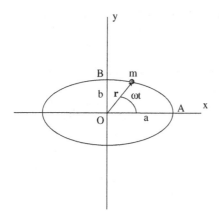

(a) Show that the particle moves on an ellipse.

(b) Show that the force acting on the particle is always directed towards the origin.

(c) Find the kinetic energy of the particle at points A and B.

(d) Find the work done by the force field in moving the particle from point A to point B by computing the appropriate line integral.

(e) Using the previous two results, show that the work done in moving from point A to point B is the kinetic energy at B minus the kinetic energy at A.

(f) Show that the total work done by the force field in moving the particle once around the ellipse is zero.

(g) Show that the force field is conservative.

(h) Find the potential energy at points A and B.

(i) Compute the work done in moving the particle from point A to point B by taking the difference in the potential energy, and show that it is the same as the answer you got in part (d) above.

5. Consider the first example in Chapter 6 (the projectile problem). Compute the work done by the net force on the projectile from the launch point to the highest point of the trajectory by:

(a) using the definition of work given by the line integral,

(b) using the definition of work given by the difference of kinetic energies at the two points.

(You should get the same answer for each.)

6. Consider Problem 1 from the Chapter 6 problems (the inclined plane problem, assuming it to have length L). Compute the work done by the

net force on the particle resulting from the motion from the top of the incline, to the bottom:

(a) using the definition of work given by the line integral,
(b) using the definition of work given by the difference of kinetic energies at the two points.

(You should get the same answer for each.)

Chapter 8

Dynamics — Conservation of Energy and Momentum

If a certain quantity associated with a system does not change in time. We say that it is *conserved*, and the system possesses a *conservation law*. Conservation laws are important since they can greatly simplify the "solution" of problems. For example, they can eliminate the need to solve differential equations in order to find the motion, or at least simplify the integration procedure (this should definitely get your attention).

The first conservation law we will study is the *law of conservation of total energy*.

Conservation of Energy for Conservative Force Fields. We consider a particle of mass m moving under the influence of a conservative force field, i.e. the force can be written as $\mathbf{F} = -\nabla V$, for some scalar valued function V. Referring to Fig. 8.1, we assume that the mass m of the particle is constant, and that at times t_1 and t_2 it is located at points P_1 and P_2, respectively, and moving with velocities $\mathbf{v}_1 = \frac{d\mathbf{r}_1}{dt}$ and $\mathbf{v}_2 = \frac{d\mathbf{r}_2}{dt}$ at points P_1 and P_2, respectively.

In Theorem 1 in the previous chapter we proved that the total work done by the force in moving the particle from P_1 to P_2 along the curve C is given by the difference in the kinetic energies at P_2 and P_1, i.e.

$$W = \frac{1}{2}mv_2^2 - \frac{1}{2}mv_1^2.$$

If the force is conservative, we proved in Theorem 2 from the previous chapter that the total work done by the force in moving the particle from P_1 to P_2 along the curve C is given by the difference in the potential energies

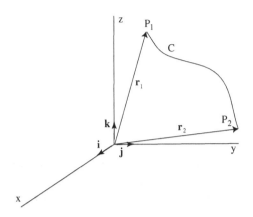

Fig. 8.1. Geometry associated with a force moving a particle between two points in space.

at P_1 and P_2, i.e.

$$W = V(P_1) - V(P_2).$$

Setting these two expressions equal gives:

$$\frac{1}{2}mv_2^2 - \frac{1}{2}mv_1^2 = V(P_1) - V(P_2),$$

or

$$\frac{1}{2}mv_2^2 + V(P_2) = \frac{1}{2}mv_1^2 + V(P_1). \tag{8.1}$$

We refer to the sum of the kinetic and potential energies of a particle as the *total energy* or just *energy* of the particle. Since the path C and the points P_1 and P_2 were completely arbitrary, (8.1) says that, under the influence of a conservative force field, a particle moves so that its total energy never changes. This is the *law of conservation of energy*, valid for conservative forces.

Constant Forces are Conservative. We have considered a number of problems concerning the motion of a particle under the influence of *only a* constant force field. We want to explicitly show that constant forces are conservative. This immediately implies that energy is conserved. We will use this fact and return to those problems and show that taking into account energy conservation greatly simplifies their solution.

Let us suppose that the constant force field has the following form:

$$\mathbf{F} = A\mathbf{i} + B\mathbf{j} + C\mathbf{k},$$

where A, B and C are constants (real numbers). In order to show that $\mathbf{F}$ is conservative we need only show that $\nabla \times \mathbf{F} = 0$ (Theorem 3 from the previous chapter). However, this should be obvious since the curl of a vector involves the partial derivatives of the vector. This vector is constant, so all the partial derivatives are zero.

Since $\mathbf{F}$ is conservative, it can be represented as the (negative) gradient of a potential function. Let us now compute the potential. However, let us simplify the problem by assuming that $\mathbf{F}$ is nonzero only in one of the coordinate directions, say $\mathbf{k}$. Then we have

$$\mathbf{F} = C\mathbf{k} = -\nabla V = -\frac{\partial V}{\partial x}\mathbf{i} - \frac{\partial V}{\partial y}\mathbf{j} - \frac{\partial V}{\partial z}\mathbf{k},$$

or

$$0 = -\frac{\partial V}{\partial x},$$

$$0 = -\frac{\partial V}{\partial y},$$

$$C = -\frac{\partial V}{\partial z}.$$

The first two of these equations imply that $V(x, y, z)$ does not depend on x or y (we might have expected this, why?). We can therefore integrate the last equation to obtain:

$$V = -Cz + c_1,$$

where c_1 is an unknown, and arbitrary, integration constant (arbitrary in the sense that its choice does not affect the force). Let us therefore make a choice. Suppose we choose $V = 0$ at $z = z_0$. Then we have:

$$0 = -Cz_0 + c_1, \quad \text{or} \quad c_1 = Cz_0,$$

and therefore:

$$V = -C(z - z_0).$$

If we consider the important case of the constant gravitational force $\mathbf{F} = -mg\mathbf{k}$, then the associated potential is:

$$V = mg(z - z_0). \tag{8.2}$$

Motion in One Dimension under a Conservative Force Field. Consider Newton's equations in one dimension where the force is given by a conservative force field, i.e.

$$m\frac{d^2s}{dt^2} = -\nabla V(s), \quad s(t_0) = s_0, \ \dot{s}(t_0) = v_0. \tag{8.3}$$

We know, since the force is conservative, that the particle moves in such a way that energy is conserved, i.e.

$$\frac{1}{2}m\dot{s}^2 + V(s) = E, \tag{8.4}$$

where E is a constant, called the *energy*. To reiterate, any solution of (9.1) must satisfy (8.4) (and vice-versa, review Chapter 5). But how are the initial conditions for a solution of (9.1) manifested in (8.4)? Through the constant E. Equation (8.4) holds for all points on a solution of (9.1), including the starting point (initial conditions). Therefore, the constant E can be expressed in terms of the (constant) initial condition of (9.1) by substituting the initial conditions of (9.1) into (8.4).

Now (8.4) can help us solve (9.1) more easily by enabling us to "skip" one integration. Here is what we mean. Equation (8.4) can be rewritten as:

$$\frac{ds}{dt} = \pm\sqrt{\frac{2}{m}}\sqrt{E - V(s)}, \tag{8.5}$$

or,

$$\pm\int_{s(t_0)}^{s(t)} \frac{ds'}{\sqrt{E - V(s')}} = \sqrt{\frac{2}{m}}\int_{t_0}^{t} dt'. \tag{8.6}$$

If we can do this integral (an "invert it") we will have solved for $s(t)$. Hence, conservation of energy reduces the solution of a *second* order ordinary differential equation to integrating a first order differential equation. Evaluating the integral depends crucially on $V(s)$ and, unfortunately, for most functions $V(s)$ we cannot do the integral analytically. Also, there is that $\pm$ sign. The proper choice there must be made when taking the square root, and depends upon the specific application under consideration.

Conservation of Momentum. While we are discussing conservation, we may as well highlight the *conservation of momentum*, although we have really already discovered it as it is expressed in Newton's laws.

Recall Newton's second law in momentum form:

$$\frac{d\mathbf{p}}{dt} = \mathbf{F}, \tag{8.7}$$

where $\mathbf{F}$ is the vector sum of all the forces. Now suppose that all of the forces are zero. Then we have:

$$\frac{d\mathbf{p}}{dt} = 0, \tag{8.8}$$

which says that the momentum is constant, in time, if there are no net forces, i.e. momentum is conserved in the absence of forces.

Now we well return to several examples and problems given earlier and solve them using conservation of energy.

Problem 3 from Chapter 5. An object of mass m is thrown vertically upward from the Earth's surface with an initial velocity $v_0\mathbf{k}$ ($v_0 > 0$). We assume that the only force acting on the object is gravity, see Fig. 8.2. Find:

1. the maximum height reached,
2. the speed as a function of its distance from the origin.

In the previous lecture we showed that the potential energy associated with the force of gravity was given by:

$$V = mg(z - z_0), \tag{8.9}$$

where z_0 is a reference position. We shall choose this so that a particle has zero potential energy at $z = 0$, i.e. we choose $z_0 = 0$.

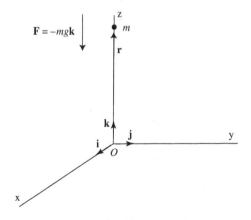

Fig. 8.2. A particle moving vertically under the influence of gravity.

With this choice, we know the total energy at $z = 0$. It is just given by the kinetic energy $\frac{1}{2}mv_0^2$. Now what characterizes the total energy at the maximum height? At the maximum height the velocity is zero, so the total energy is given by the potential energy mgz_{max}. Thus, equating the total energy at $z = 0$ to the total energy at z_{max} gives:

$$\frac{1}{2}mv_0^2 = mgz_{max}, \quad \text{or} \quad z_{max} = \frac{v_0^2}{2g}.$$

How would we calculate the velocity at an arbitrary height? Well, energy is the same at *any* height since it is conserved. The energy at an arbitrary height z is given by:

$$E_z = \frac{1}{2}mv^2 + mgz.$$

This must be equal to the energy at $z = 0$, which is just the kinetic energy at $z = 0$:

$$E_{z=0} = \frac{1}{2}mv_0^2.$$

Since energy is conserved $E_{z=0} = E_z$, and therefore:

$$\frac{1}{2}mv_0^2 = \frac{1}{2}mv^2 + mgz, \quad \text{or} \quad v = \sqrt{v_0^2 - 2gz}.$$

Since $v = \frac{dz}{dt}$ we can also solve for the **height as a function of time** by integrating the equation for $v = \frac{dz}{dt}$, i.e.

$$v = \frac{dz}{dt} = \sqrt{v_0^2 - 2gz},$$

or

$$\int_0^z \frac{dz'}{\sqrt{v_0^2 - 2gz'}} = \left. \frac{-2\sqrt{v_0^2 - 2gz'}}{2g} \right|_0^z = \int_0^t dt' = t.$$

After a bit of algebra you obtain:

$$z = v_0 t - \frac{1}{2}gt^2.$$

Revisiting an example from Chapter 6. A particle, denoted by P, sits at the top (point A in the figure) of a frictionless fixed sphere of radius b, see Fig. 8.3.

The particle is displaced slightly so that it slides (without rolling) down the sphere.

Compute:

1. the position of the particle as it leaves the sphere,
2. the speed of the particle at the instant it leaves the sphere.

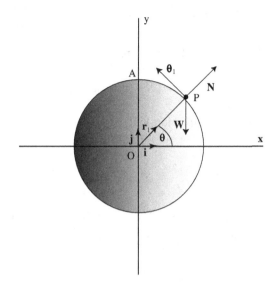

Fig. 8.3. A particle moving on a sphere.

First, the potential energy due to the gravitational force is given by $V = mg(y - y_0)$, and we take the reference point to be $y_0 = 0$, meaning that the potential energy is chosen to be zero at $y = 0$.

Using conservation of energy, we have:

$$\text{P.E. at A} + \text{K. E. at A} = \text{P.E. at P} + \text{K. E. at P}$$
$$mgb \quad + \quad 0 \quad = mgb\sin\theta + \quad \tfrac{1}{2}mv^2,$$

From which it follows that:

$$v^2 = 2gb(1 - \sin\theta). \tag{8.10}$$

But remember here that $v = b\dot{\theta}$.

Now we have to ask ourselves, "what characterizes the particle leaving the sphere?" We argued in Chapter 4 that this was when the normal reaction force was zero. This means that we will not be able to avoid Newton's equations entirely.

Recall from Chapter 3 that the acceleration of a particle constrained to move along a circle is given by:

$$\mathbf{a} = -\frac{v^2}{b}\mathbf{r}_1 + \frac{dv}{dt}\boldsymbol{\theta}_1.$$

Hence, Newton's equations are given by:

$$m\left(-\frac{v^2}{b}\mathbf{r}_1 + \frac{dv}{dt}\boldsymbol{\theta}_1\right) = \mathbf{W} + \mathbf{N},$$

$$= (N - mg\sin\theta)\mathbf{r}_1 - mg\cos\theta\boldsymbol{\theta}_1.$$

The $\mathbf{r}_1$ component of this equation will give us a relationship between the magnitude of the normal force, and θ and v:

$$-m\frac{v^2}{b} = N - mg\sin\theta. \tag{8.11}$$

Now setting $N = 0$ in (8.11) gives us a relationship between the value of v^2 and θ for when the particle leaves the sphere:

$$v^2 = bg\sin\theta.$$

Equation (8.10) gives the value of v^2 for any angle θ. Hence, equating these two expressions gives us an equation for the angle at which the particle leaves the sphere:

$$2gb(1 - \sin\theta) = bg\sin\theta, \quad \text{or} \quad \sin\theta = \frac{2}{3}.$$

The speed at which the particle leaves the sphere is found by substituting this value into (8.10):

$$v^2 = b^2\dot{\theta}^2 = \frac{2}{3}gb, \quad \text{or} \quad |\dot{\theta}| = \sqrt{\frac{2g}{3b}}.$$

Problem Set 8

1. Consider a projectile of constant mass m that is launched with initial speed v_0 at an angle α with the horizontal. Using conservation of energy, compute:

(a) the maximum height reached,
(b) the position vector at any time.

For an illustration of the geometry see the following figure.

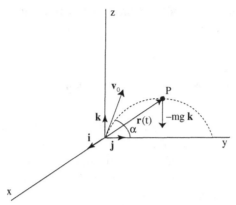

2. A particle P of constant mass m slides without rolling down a frictionless inclined plane of angle α and length ℓ, see the following figure.

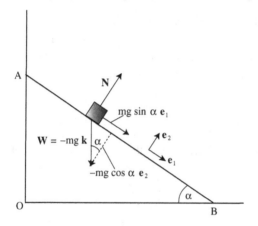

If the particle starts from rest at the top of the incline (at point A) use conservation of energy to determine the following quantities:

(a) the velocity as a function of distance traveled,

(b) the distance traveled after time t.

(c) Compute the magnitude of the acceleration.

3. Consider a particle of mass m moving with velocity v_0 and a particle of mass M moving with velocity V_0, where both particles are constrained to move along a line and there is no net force acting on the particles, see the following figure. (Note that since the motion is one-dimensional, we are dispensing with the use of unit vectors in expressing the vectorial

nature of velocity — the sign of the quantity is adequate for determining the direction.) We assume that the two particles collide.

(a) Relate the momentum of the two particles before the collision to the momentum of the two particles after the collision.

(b) Let x denote the position of the particle of mass m and let X denote the position of the particle of mass M. We define the *center of mass* of the system of two particles by:

$$\xi = \frac{mx + MX}{m + M}.$$

Show that the velocity of the center of mass, i.e. $\dot{\xi}$, is constant.

(c) An *elastic collision* is defined as a collision where the sum of the kinetic energy of the two particles is the same before and after the collision, i.e.

$$\frac{1}{2}mv_0^2 + \frac{1}{2}MV_0^2 = \frac{1}{2}mv^2 + \frac{1}{2}MV^2. \tag{8.12}$$

Using this result, and the result from part (a), show that the velocities before and after the collision are related as follows:

$$v = \frac{m - M}{m + M}v_0 + \frac{2M}{m + M}V_0,$$

$$V = \frac{M - m}{m + M}V_0 + \frac{2m}{m + M}v_0. \tag{8.13}$$

(d) Consider the case of an elastic collision with $m = M$. What are v and V?

(e) Consider the case of an elastic collision with M "very large compared to m". Determine v and V in general, and describe the motion in the special case that $V_0 = 0$.

Chapter 9

Dynamics — The Phase Plane for One-Dimensional Motion

We saw in Chapter 8 that a solution to Newton's equation in one dimension *with conservative forces*, i.e.

$$m\frac{d^2s}{dt^2} = -\nabla V(s), \quad s(t_0) = s_0, \ \dot{s}(t_0) = v_0, \tag{9.1}$$

must satisfy

$$\frac{1}{2}m\dot{s}^2 + V(s) = E, \tag{9.2}$$

where E is a constant, called the *total energy*, which is determined by the initial conditions $s(t_0) = s_0, \dot{s}(t_0) = v_0$.[1] We now want to exploit this fact and develop a graphical way of understanding *all* possible solutions to (9.1) (i.e. for any choice of initial condition). This technique is called *phase plane analysis*. It works for differential equations in general, and you will learn more about that later on.

First, what is this phase "plane"? After all, we are studying motion in one dimension. The first step to understanding this is a closer examination of (9.2) and giving it a *geometrical interpretation*. Let us consider (9.2) as a function of two variables (which it is, of course), s and $\dot{s}$ (and, you guessed it, the *phase plane* will be the plane with coordinate axes s and $\dot{s}$).

---

[1]Note that in one dimension the gradient is a scalar (as opposed to a vector in two or three dimensions) and has the much simpler form, $\nabla V(s) \equiv \frac{dV}{ds}(s)$.

We will call this function:

$$H(s,v) = \frac{1}{2}mv^2 + V(s), \tag{9.3}$$

where we now denote

$$\dot{s} \equiv v.$$

We refer to $H(s,v)$ as the *energy function*, and the interpretation we have from above is that the *level sets, or level curves* of the energy function are trajectories of (9.1) (we will return to the initial condition question shortly, but, for the moment, we will leave it as we focus on developing this geometrical picture). The *level sets, or level curves* of a function are the set of points in the domain of a function where the function is a constant. In our case, the level sets of $H(s,v)$ are the sets of points in the $s-v$ plane for which $H(s,v) = $ constant. "Typically", these level sets are curves. Let us consider a familiar example.

Example 12. Consider the function:

$$H(s,v) = s^2 + v^2.$$

The level sets of this function, i.e.

$$\{(s,v) \in \mathbb{R}^2 | H(s,v) = s^2 + v^2 = E = \text{constant}\},$$

are the circles of radius $\sqrt{E}$. Clearly, the level sets are defined only for constants $E \geq 0$. They are circles (curves) for $E > 0$. For $E = 0$ the level set is a point, the origin. For certain exceptional values the level sets may degenerate to a point (which is why we put "typically" in quotes above).

So now you see how to "solve" (9.1). Write down the energy function and plot its level sets in the $s-v$ plane. But in what sense does this "solve" (9.1)? We need to develop this geometrical picture further in order to see this.

Suppose $s(t)$ is a solution of (9.1). Then it follows that $(s(t), \dot{s}(t) \equiv v(t))$ lies on a level set of the energy function, for some energy (constant) E, i.e.

$$H(s(t), v(t)) = \frac{1}{2}mv(t)^2 + V(s(t)) = E. \tag{9.4}$$

Then we have:

$$\frac{d}{dt}H(s(t), v(t)) = \frac{\partial H}{\partial s}\frac{ds}{dt} + \frac{\partial H}{\partial v}\frac{dv}{dt} = \nabla H \cdot \left(\frac{ds}{dt}, \frac{dv}{dt}\right) = 0. \tag{9.5}$$

As t varies, $(s(t), v(t))$ traces out a curve in the phase plane, which is a solution of (9.1), and a level set of (9.2). Now, for any fixed t

$\left(\frac{ds}{dt}(t), \frac{dv}{dt}(t)\right)$ is a vector tangent to the level set. It then follows from (9.5) that $\left(\frac{\partial H}{\partial s}(s(t), v(t)), \frac{\partial H}{\partial v}(s(t), v(t))\right) = \nabla H(s(t), v(t))$ is a vector perpendicular to the level set at that point (in the sense of being perpendicular to the tangent vector to the curve at that point).

For each value of t, $\left(\frac{ds}{dt}(t), \frac{dv}{dt}(t)\right)$ is a vector in the $s - v$ plane that is tangent to a trajectory of (9.1). It is in this sense that we say that (9.1) defines a *vector field* in the phase plane, and the solutions of (9.1) are curves, parametrised by t, have the property that for each t the tangent vector is given by the vector field. This is made more explicit if we rewrite (9.1) in the form of a first order equation for the vector $\left(\frac{ds}{dt}(t), \frac{dv}{dt}(t)\right)$. This is done as follows:

$$\dot{s} = v,$$

$$\ddot{s} = \dot{v} = -\frac{1}{m}\nabla V(s). \tag{9.6}$$

It should be clear that (9.6) is equivalent to (9.1) in the sense that a solution of (9.6) is equivalent to a solution of (9.1), and vice-versa. The geometrical picture that we have developed is illustrated in Fig. 9.1.

Now we describe a graphical way of drawing the level sets of the energy function in the phase plane, or *constructing the phase portrait*. We can solve (8.4) for v as a function of E and s:

$$v = \pm\sqrt{\frac{2}{m}}\sqrt{E - V(s)}. \tag{9.7}$$

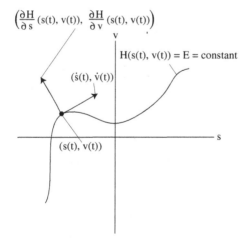

Fig. 9.1. Geometry of motion associated with a level set of the energy.

For fixed E, this defines the corresponding level set of the energy function as the graph of a function of s. Well, "almost", there is the $\pm$ to consider. However, this actually makes things simpler. We must deal with both the $+$ and $-$ signs separately. First we consider the $+$ sign. Since the square root of a function is positive then v will be positive. We then plot the graph of the function. Next we consider the minus sign. *But it is the same graph, with just an overall minus sign.* Hence, we just reflect the graph constructed with the plus sign about the horizontal axis in the phase plane (you should think carefully about this!).

But still, how do we construct the graph to begin with, with the plus sign? There is a simple graphical procedure for this, which we now describe.

Plot the potential energy as a function of s, and immediately below this graph draw the s and v axes of the phase plane, as shown in Fig. 9.2.

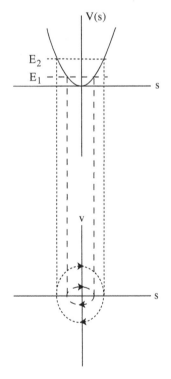

Fig. 9.2. Graphical procedure for drawing the phase portrait from the potential energy.

For the potential energy function sketched in the figure, we will sketch the phase portrait. This means sketch the level sets of the energy function (8.4) for different values of E.

Pick a value of E, say E_1. Draw the horizontal line $V(s) = E_1$ on the graph of the potential as shown in Fig. 9.2. Now consider (9.7) with the plus sign. From this expression it should be clear that v exists only for the range of s for which $E_1 - V(s) \geq 0$. The values of s for which $E_1 - V(s) = 0$ (i.e. $v = 0$) are called *turning points*. Start at the left hand value of s for which $E_1 - V(s) = 0$. Then move to the right (i.e. increase s). In this case v increases since $E_1 - V(s)$ increases. As we move to the right, $E_1 - V(s)$ continues to increase until it reaches a maximum value, then it decreases to zero at the right-hand point where $E_1 - V(s) = 0$. This results in a curve in the positive v half plane. If we take the minus sign in (9.7), we get the negative of this curve, which results in the closed curve shown in Fig. 9.2. We see that the turning points are aptly named since the velocity changes direction at those points. We have also shown a direction of motion on these curves. You should convince yourself that this is correct. We can repeat this construction for a larger value of energy, say E_2 as shown in Fig. 9.2. We get another closed path that contains the one previously constructed with lower energy.

What about the point at the origin, which is a local minimum of the potential? At this point $\frac{dV}{ds}(0) = 0$, which implies that $\dot{v} = 0$ from (9.6). Also, $v = 0$, which implies $\dot{s} = 0$ from (9.6). Hence, the vector field is zero at this point. The point cannot "move". It is called an *equilibrium point*. In general, any point for which (s, v), where $v = 0$ and s satisfies $\frac{dV}{ds} = 0$ is an equilibrium point. For this example we say that it is a *stable equilibrium point* since solutions starting nearby oscillate around it, i.e. they never move far away. The closed curves in Fig. 9.2 are examples of periodic solutions.

Now let us consider a more complicated example, as shown in Fig. 9.3. What makes this example more complicated is that the potential has multiple *relative extrema* (sometimes called "critical points").

First, note that there are three equilibrium points, two stable (corresponding to relative minima of the potential), and one *saddle point* (corresponding to a relative maximum of the potential). Following the graphical procedure described above, we obtain the level sets of the energy function for the four different values of energy shown in Fig. 9.3.

You should be able to convince yourself that for energies E_1, E_2, and E_4 you obtain closed curves in the phase plane (but think about where

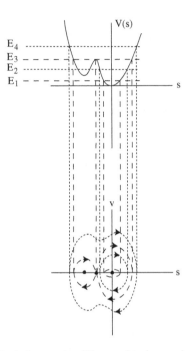

Fig. 9.3. A more complicated example of the graphical procedure for drawing the phase portrait from the potential energy.

the "dimples" in the closed curves arise from). The energy E_3 is slightly different. This is the energy of the saddle point, and now you will be able to see where the name comes from. At this energy a level set of the energy function passes through this point as shown in Fig. 9.3. If you take into consideration the direction of motion of the level set, the saddle point nature is clear, but let us look at this point more closely.

For Figs. 9.2 and 9.3 why did we put the arrows on the curves as indicated (i.e. I am worried about getting the direction right)? The arrows are in the direction of motion. Refer back to (9.6). Note that for $v > 0$, we have $\dot{s} > 0$, which means that the direction is from left to right. For $v < 0$, we have $\dot{s} < 0$, which means that the direction is from right to left. Make sure you understand this argument.

In Fig. 9.3 the level set of the energy function with energy E_3 is an example of a *separatrix*, i.e. a curve that separates qualitatively distinct level sets of the energy function. It is the level set that passes through a relative maximum of the potential energy.

The phase portrait is understood if we sketch the *qualitatively distinct* curves in the phase plane? What do we mean by qualitatively distinct? In Fig. 9.2 all of the level sets of the energy function were "concentric" closed curves in the phase plane. We say that they are all qualitatively similar. Each could be smoothly deformed into any other by shrinking or growing, and without passing through any equilibria. That is not the case in Fig. 9.3 due to the presence of the saddle point and its associated separatrix. The separatrix defines the boundary between regions with qualitatively similar level sets of the energy function. Therefore, if there are no saddle points the phase portrait is fairly simple.

There have been a lot of ideas in this chapter. Now we summarize the main points.

Phase Plane. The plane with horizontal axis given by s, and vertical axis given by $\dot{s} = v$.

Level Set of the Energy Function. The set of points in the phase plane with equal energy. Typically, this is one or more curves, but could also be an isolated point. These are sometimes called *phase curves, or trajectories*.

Closed Level Set of the Energy Function, not Containing any Equilibria. These are periodic solutions of the system.

Turning Point. The point on the horizontal axis where $\dot{v} = 0$.

Equilibrium Point. A point in the phase plane where $\dot{s} = 0$ *and* $\dot{v} = 0$, simultaneously. Equivalently, a relative extrema of the potential energy.

Stable Equilbrium Point. An equilibrium point which is a relative minima of the potential energy, i.e. at the equilibrium point we have $\frac{dV}{ds} = 0$ and $\frac{d^2V}{ds^2} > 0$.

Saddle Point. An equilibrium point which is a relative maxima of the potential energy, i.e. at the equilibrium point we have $\frac{dV}{ds} = 0$ and $\frac{d^2V}{ds^2} < 0$.

Phase Portrait. An illustration of the qualitatively distinct level sets of the energy function in the phase plane.

Separatrix. A level set of the energy function that passes through a relative maximum of the potential energy. It separates qualitatively distinct level sets of the energy function.

Vector Field. The field of velocity vectors in the phase plane defined by the two-dimensional, first order form of Newton's equations.

Problem Set 9

1. Consider the equation:

$$\ddot{s} = -s,$$

where, for simplicity, we have set $m = 1$ (remember from an earlier homework problem that we can "rescale time" so that the mass becomes unity).

(a) Write it as a first order system, or vector field, on the phase plane.
(b) Compute the potential energy and sketch it.
(c) Find all equilibria and classify their stability.
(d) Sketch the phase portrait.
(e) Compute expressions for the trajectories in the phase plane as a function of time (and take $s(0) = 0$ for simplicity).

2. Consider the equation:

$$\ddot{s} = s - s^3,$$

where, for simplicity, we have set $m = 1$.

(a) Write it as a first order system, or vector field, on the phase plane.
(b) Compute the potential energy and sketch it.
(c) Find all equilibria and classify their stability.
(d) Sketch the phase portrait.

3. Consider the equation:

$$\ddot{s} = s - s^2,$$

where, for simplicity, we have set $m = 1$.

(a) Write it as a first order system, or vector field, on the phase plane.
(b) Compute the potential energy and sketch it.
(c) Find all equilibria and classify their stability.
(d) Sketch the phase portrait.

Chapter 10

Dynamics — The Simple Pendulum. Torque and Angular Momentum

We consider the classic pendulum, a particle of mass m connected to some point by an inextensible (and massless) connector ("string") of constant length ℓ, which is free to move in a circle of radius ℓ in a vertical plane. We assume that the only force acting on the mass is gravity, see Fig. 10.1.

First, we will derive Newton's equation for the pendulum. But we need to take a step back and first address some kinematical considerations (i.e. how we will describe the motion). We will consider the motion to be one-dimensional with the mass being constrained to lie on a circle of radius ℓ. Therefore it is natural to describe the location of the mass by an angle, say ϑ, which we take as the deviation from vertical (i.e. the pendulum hanging straight down).

We will derive the equations of motion two ways: (1) by the energy method, and (2) by writing down Newton's second Law (i.e. "Newton's equations").

The Energy Method

The only forces acting on the pendulum mass (that we are considering) are gravity, and we know (make sure you do know this!) that gravity is a conservative force. Hence, the total energy, kinetic plus potential energy, is conserved.

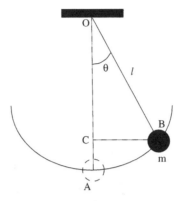

Fig. 10.1. Geometry associated with the pendulum.

The potential energy is solely due to gravity, and is given by mgh, where h is the height above a reference position. We will take this position (remember, it can be chosen arbitrarily without affecting the equations of motion) to be the position where the pendulum is at its lowest point, i.e. "hanging straight down". Using a little bit of trigonometry, the height above this position, as a function of θ, is given by $\ell(1-\cos\theta)$ (see Fig. 10.1). Therefore the potential energy is given by $V = mg\ell(1 - \cos\theta)$.

Now focus on the kinetic energy. It is certainly one half times the mass times the square of the velocity. But what is the velocity? It is the time derivative of the position vector. The position vector is given by:

$$\mathbf{r} = \ell\mathbf{r}_1,$$

and therefore

$$\dot{\mathbf{r}} = \ell\dot{\mathbf{r}}_1 = \ell\dot{\theta}\boldsymbol{\theta}_1.$$

So the total energy is:

$$\frac{1}{2}m\dot{\mathbf{r}}\cdot\dot{\mathbf{r}} + V(\theta) = E,$$

or

$$\frac{1}{2}m\ell^2\dot{\theta}^2 + mg\ell(1 - \cos\theta) = E.$$

Now differentiate this expression with respect to time:

$$m\ell^2\ddot{\theta}\dot{\theta} + mg\ell\sin\theta\dot{\theta} = 0,$$

or, after cancelling $\dot{\theta}$ and rearranging terms:

$$\ddot{\theta} = -\frac{g}{\ell}\sin\theta.$$

Newton's Equations

Newton's equations are given by:

$$m\ddot{\mathbf{r}} = \mathbf{F}.$$

The mass is constrained to lie on a circle. So we are only interested in the acceleration and forces in the tangential direction, i.e. $\boldsymbol{\theta}_1$. Recall from above that:

$$\mathbf{r} = \ell \mathbf{r}_1,$$

from which it follows, after differentiating with respect to time, that the component of acceleration in the $\boldsymbol{\theta}_1$ direction is given by:

$$\ell \ddot{\theta} \boldsymbol{\theta}_1.$$

The gravitational force is given by $-mg\mathbf{k}$. The component of this force in the $\boldsymbol{\theta}_1$ direction is given by:

$$(-mg\mathbf{k} \cdot \boldsymbol{\theta}_1)\, \boldsymbol{\theta}_1 = -mg \cos\left(\frac{\pi}{2} - \theta\right) \boldsymbol{\theta}_1 = -mg \sin\theta \boldsymbol{\theta}_1.$$

From these two expressions we obtain:

$$m\ell \ddot{\theta} = -mg \sin\theta,$$

or

$$\ddot{\theta} = -\frac{g}{\ell} \sin\theta. \tag{10.1}$$

Now we want to sketch the phase portrait for the pendulum. However, the presence of the constants g and ℓ present a small problem in that we would wonder if the phase portrait would change significantly if different values of these constants were taken. We can deal with this by eliminating them completely through *rescaling time* and making the resulting equation *dimensionless*. First, note that $\sqrt{\frac{g}{\ell}}$ has the dimensions of $\frac{1}{\text{time}}$. We can therefore define a *dimensionless time*, called, τ, as follows:

$$t = \sqrt{\frac{\ell}{g}}\tau.$$

Then

$$\frac{d^2}{dt^2} = \frac{g}{\ell}\frac{d^2}{d\tau^2},$$

and, in the dimensionless time τ, (10.1) becomes:

$$\frac{d^2\theta}{d\tau^2} = -\sin\theta. \tag{10.2}$$

In dimensionless form, we take as the potential energy:

$$V(\theta) = 1 - \cos\theta,$$

since, using this form for $V(\theta)$, (10.2) can be written as:

$$\frac{d^2\theta}{d\tau^2} = -\frac{dV}{d\theta}(\theta).$$

In Fig. 10.2 we plot the potential energy and the phase portrait using the graphical procedure as described in Chapter 9.

It is important to be aware that the phase plane is periodic in θ. What does this mean in terms of the motion of the pendulum? There are two

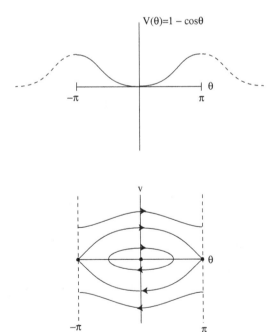

Fig. 10.2. The potential energy and the phase portrait associated with the pendulum.

equilibrium points: a stable equilibrium at $\theta = 0$, and a saddle at $\theta = \pi$. There are two separatrices that pass through the saddle point. The separatrices bound a family of closed level sets of the energy function (periodic solutions of Newton's equations) periodic solutions of Newton's equations. They have the property that the angle variable never goes beyond $\pm\pi$. Can you give a physical description of such periodic solutions? They are called *librations*, or *librational motions*. The periodic solutions above and below the separatrices are called *rotations*. Why?

Dynamics–Torque and Angular Momentum

Consider a particle moving under the action of a force **F**. Let **r** denote the position vector of the particle with respect to a point fixed in space, which we call O, and take as the origin of a cartesian cordinate system, see Fig. 10.3.

We define the *torque about O*, or the *moment of the force F about O* as:

$$\mathbf{\Lambda} = \mathbf{r} \times \mathbf{F}. \qquad (10.3)$$

You can think of the magnitude of the torque as a measure of the "turning effect" of the force on the particle *about the chosen point O*.

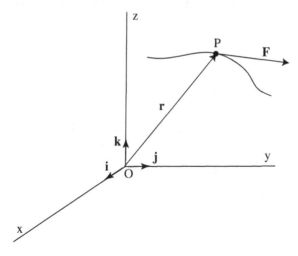

Fig. 10.3. Geometry associated with a force acting on a particle moving in three-dimensional space.

Theorem 5.

$$\mathbf{r} \times \mathbf{F} = \frac{d}{dt}\left(m(\mathbf{r} \times \mathbf{v})\right).$$

Proof. This is just a straightforward computation.

$$\frac{d}{dt}\left(m(\mathbf{r} \times \mathbf{v})\right) = m\left(\frac{d\mathbf{r}}{dt} \times \mathbf{v} + \mathbf{r} \times \frac{d\mathbf{v}}{dt}\right) = m(\mathbf{v} \times \mathbf{v}) + \mathbf{r} \times m\frac{d\mathbf{v}}{dt}$$

$$= \mathbf{r} \times \mathbf{F} = \mathbf{\Lambda}. \qquad \square$$

The quantity:

$$\mathbf{\Omega} = m(\mathbf{r} \times \mathbf{v}) = \mathbf{r} \times \mathbf{p}, \qquad (10.4)$$

is called the *angular momentum about O*, or the *moment of momentum about O*.

The theorem states the torque acting on a particle (about a point O) is equal to the time rate of change of its angular momentum (about the same point O), i.e.

$$\mathbf{\Lambda} = \frac{d\mathbf{\Omega}}{dt}. \qquad (10.5)$$

Just as linear momentum is conserved if there are no net exernal forces on a particle, angular momentum about a point is conserved if there are no net torques on the particle about the same point. To see this, set $\mathbf{\Lambda} = 0$ in (10.5). This gives:

$$\frac{d}{dt}\left(m(\mathbf{r} \times \mathbf{v})\right) = 0, \qquad (10.6)$$

or

$$m(\mathbf{r} \times \mathbf{v}) = \text{constant}. \qquad (10.7)$$

Example 13. Recall the problem from Chapter 7 problem set. A particle of mass m moves in the $x - y$ plane so that its position vector is:

$$\mathbf{r} = a\cos\omega t\mathbf{i} + b\sin\omega t\mathbf{j},$$

where a, b, and ω are positive constants with $a > b$. Here we consider questions related to angular momentum and torque on the particle. In particular, we will:

1. compute the torque on the particle about the origin,
2. compute the angular momentum of the particle about the origin,
3. show that angular momentum is conserved.

In the problem we showed previously that the force on the particle is given by:

$$\mathbf{F} = -m\omega^2\mathbf{r}.$$

Therefore, the torque on the particle about the origin due to this force is given by:

$$\mathbf{\Lambda} = \mathbf{r} \times \mathbf{F} = \mathbf{r} \times (-m\omega^2\mathbf{r}) = -m\omega^2(\mathbf{r} \times \mathbf{r}) = 0.$$

Furthermore, it was shown in the problem previously that the velocity of the particle is given by:

$$\mathbf{v} = -\omega a \sin\omega t\mathbf{i} + \omega b \cos\omega t\mathbf{j},$$

and therefore the momentum is given by:

$$\mathbf{p} = -m\omega a \sin\omega t\mathbf{i} + m\omega b \cos\omega t\mathbf{j}.$$

Then the angular momentum about O is given by:

$$\mathbf{\Omega} = \mathbf{r} \times \mathbf{p} = \begin{vmatrix} \mathbf{i} & \mathbf{j} & \mathbf{k} \\ a\cos\omega t & b\sin\omega t & 0 \\ -m\omega a\sin\omega t & m\omega b\cos\omega t & 0 \end{vmatrix} = m\omega ab\mathbf{k}.$$

Clearly, the angular momentum about O is constant, but we knew that should be the case since the torque about O is zero.

Problem Set 10

Problems 1–7 below are concerned with the motion of a particle of constant mass m in one dimension under the action of a conservative force:

$$\dot{s} = v,$$

$$\dot{v} = -\frac{1}{m}\frac{dV}{ds}(s),$$

where $V(s)$ is the potential function.

1. Show that all level sets of the energy function cross the s-axis in the phase plane perpendicularly. (Hint: Compute the slope of the level set at the turning point, $v = 0$.)
2. Suppose you add a constant (i.e. a real number) to the potential function. How does the corresponding phase portrait change?
3. Suppose $(s, v) \equiv (s_0, 0)$ is an equilibrium point of Newton's equations above. Is it a solution of Newton's equations?

4. Consider the potential energy sketched in figure below.

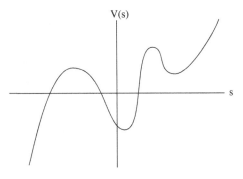

(a) Determine the number of equilibria and their stability type.
(b) Sketch the phase portrait.

5. Consider the potential energy sketched in figure below.

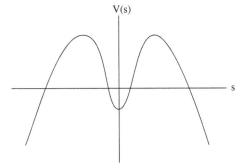

(a) Determine the number of equilibria and their stability type.
(b) Sketch the phase portrait.

6. Consider the potential energy sketched in figure below.

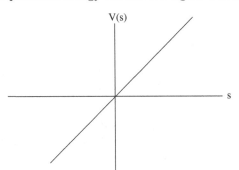

(a) Determine the number of equilibria and their stability type.
(b) Sketch the phase portrait.

7. Consider the potential energy sketched in figure below.

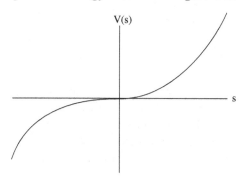

There is an inflection point at the origin (i.e. think of $V(s) = s^3$).

(a) Determine the number of equilibria and their stability type.
(b) Sketch the phase portrait.

8. The angular momentum (about some point) as a function of time is given by:

$$\mathbf{\Omega} = 6t^2\mathbf{i} - (2t + 1)\mathbf{j} + (12t^3 - 8t^2)\mathbf{k}.$$

Find the torque (about the same point) at $t = 1$.

9. A particle of mass 2 moves in a time-dependent force field given by:

$$\mathbf{F} = 24t^2\mathbf{i} + (36t - 16)\mathbf{j} - 12t\mathbf{k}.$$

Assume that at $t = 0$ the particle has the following position and velocity:

$$\mathbf{r}_0 = 3\mathbf{i} - \mathbf{j} + 4\mathbf{k},$$

$$\mathbf{v}_0 = 6\mathbf{i} + 15\mathbf{j} - 8\mathbf{k}.$$

Compute, for any time t:

(a) the velocity,
(b) the position,
(c) the torque about the origin,
(d) the angular momentum about the origin.

10. A *central force* is a force having the following form:

$$\mathbf{F} = f(r)\mathbf{r_1} = f(r)\frac{\mathbf{r}}{r},$$

where $f(r)$ is an arbitrary function of the magnitude of the position vector. Show that angular momentum about the origin is conserved for a particle of constant mass m moving under the influence of a central force field.[1]

11. Show that if a particle moves under the influence of a central force field, then its path must always lie in a plane. (Hint. A plane is defined by a constant vector. It then suffices to show that the position vector is perpendicular to an appropriately chosen constant vector. Use the previous problem to choose this constant vector.)[2]

---

[1]This is an important result that we will use in the next chapter.

[2]This is also an important result that we will use in the next chapter.

Chapter 11

Motion in a Central Force Field

In this chapter we are going to study the properties of a particle of (constant) mass m moving in a particular type of force field, a *central force field*. Central forces are very important in physics and engineering. For example, the gravitational force of attraction between two point masses is a central force. The Coulomb force of attraction and repulsion between charged particles is a central force. Because of their importance they deserve special consideration. We begin by giving a precise definition of *central force, or central force field*.

Central Forces: The Definition. Suppose that a force acting on a particle of mass m has the properties that:

- the force is always directed from m toward, or away, from a fixed point O,
- the magnitude of the force only depends on the distance r from O.

Forces having these properties are called *central forces*. The particle is said to move in a *central force field*. The point O is referred to as the *center of force*.

Mathematically, $\mathbf{F}$ is a central force if and only if:

$$\mathbf{F} = f(r)\mathbf{r}_1 = f(r)\frac{\mathbf{r}}{r}, \qquad (11.1)$$

where $\mathbf{r}_1 = \frac{\mathbf{r}}{r}$ is a unit vector in the direction of $\mathbf{r}$.

If $f(r) < 0$ the force is said to be *attractive towards O*. If $f(r) > 0$ the force is said to be *repulsive from O*. We give a geometrical illustration in Fig. 11.1.

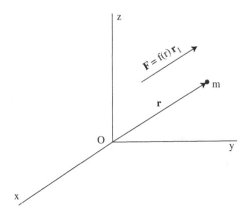

Fig. 11.1. Geometrical illustration of a central force.

Properties of a Particle Moving under the Influence of a Central Force. If a particle moves in a central force field then the following properties hold:

1. The path of the particle must be a *plane curve*, i.e. it must lie in a plane. (This was a problem in Chapter 10.)
2. The angular momentum of the particle is conserved, i.e. it is constant in time. (This was a problem in Chapter 10.)
3. The particle moves in such a way that the position vector (from the point O) sweeps out equal areas in equal times. In other words, the time rate of change in area is constant. This is referred to as the *Law of Areas.* We will describe this in more detail, and prove it, shortly.

Equations of Motion for a Particle in a Central Force Field. Now we will derive the basic equations of motion for a particle moving in a central force field.

From Property 1 above, the motion of the particle must occur in a plane, which we take as the xy plane, and the center of force is taken as the origin. In Fig. 11.2 we show the xy plane, as well as the polar coordinate system in the plane.

Since the vectorial nature of the central force is expressed in terms of a radial vector from the origin it is most natural (though not required!) to write the equations of motion in polar coordinates. In earlier lectures we derived the expression for the acceleration of a particle in polar coordinates:

$$\mathbf{a} = (\ddot{r} - r\dot{\theta}^2)\mathbf{r}_1 + (r\ddot{\theta} + 2\dot{r}\dot{\theta})\boldsymbol{\theta}_1. \tag{11.2}$$

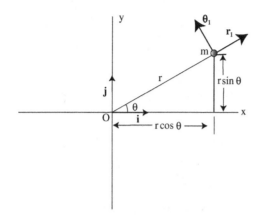

Fig. 11.2. Polar coordinate system associated with a particle moving in the xy plane.

Then, using Newton's second law, and the mathematical form for the central force given in (11.1), we have:

$$m(\ddot{r} - r\dot{\theta}^2)\mathbf{r}_1 + m(r\ddot{\theta} + 2\dot{r}\dot{\theta})\boldsymbol{\theta}_1 = f(r)\mathbf{r}_1, \tag{11.3}$$

or

$$m(\ddot{r} - r\dot{\theta}^2) = f(r), \tag{11.4}$$

$$m(r\ddot{\theta} + 2\dot{r}\dot{\theta}) = 0. \tag{11.5}$$

These are the basic equations of motion for a particle in a central force field. They will be the starting point for many of our investigations.

From these equations we can derive a useful *constant of the motion*. This is done as follows. From (11.5) we have:

$$m(r\ddot{\theta} + 2\dot{r}\dot{\theta}) = \frac{m}{r}(r^2\ddot{\theta} + 2r\dot{r}\dot{\theta}) = \frac{m}{r}\frac{d}{dt}(r^2\dot{\theta}) = 0,$$

or

$$r^2\dot{\theta} = \text{constant} = h. \tag{11.6}$$

This is an interesting relation that, we will see, is related to properties 2 and 3 above. However, one use for it should be apparent. If you know the r component of the motion it allows you to compute the θ component by integration. This is another example of how *constants of the motion* allow us to "integrate" the equations of motion. It also explain why constants of the motion are often referred to as *integrals of the motion*.

Now, let us return to property 3 above and derive the *Law of Areas*.

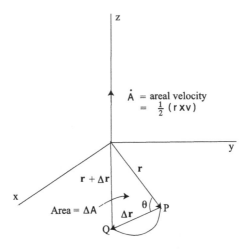

Fig. 11.3. Illustration of the Law of Areas.

Suppose that in time Δt the position vector moves from $\mathbf{r}$ to $\mathbf{r} + \Delta \mathbf{r}$. Then the area swept out by the position vector in this time is approximately half the area of a parallelogram with sides $\mathbf{r}$ and $\Delta \mathbf{r}$. We give a proof of this:

$$\text{Area of parallelogram} = \text{height} \times |\mathbf{r}|,$$

$$= |\Delta \mathbf{r}| \sin \theta |\mathbf{r}|,$$

$$= |\mathbf{r} \times \Delta \mathbf{r}|,$$

see Fig. 11.3.

Hence,

$$\Delta A = \frac{1}{2} |\mathbf{r} \times \Delta \mathbf{r}|.$$

Dividing this expression by Δt, and letting $\Delta t \to 0$, gives:

$$\lim_{\Delta t \to 0} \frac{\Delta A}{\Delta t} = \lim_{\Delta t \to 0} \frac{1}{2} \left| \mathbf{r} \times \frac{\Delta \mathbf{r}}{\Delta t} \right| = \frac{1}{2} |\mathbf{r} \times \mathbf{v}|,$$

or

$$\dot{A} = \frac{1}{2} |\mathbf{r} \times \mathbf{v}|.$$

Now we need to evaluate $\mathbf{r} \times \mathbf{v}$. Using $\mathbf{r} = r\mathbf{r}_1$, we have:

$$\mathbf{r} \times \mathbf{v} = \mathbf{r} \times (\dot{r}\mathbf{r}_1 + r\dot{\theta}\boldsymbol{\theta}_1) = \dot{r}(\mathbf{r} \times \mathbf{r}_1) + r\dot{\theta}(\mathbf{r} \times \boldsymbol{\theta}_1) = r^2\dot{\theta}\mathbf{k}.$$

Therefore, we have:

$$r^2\dot{\theta} = 2\dot{A} = \text{constant.} \tag{11.7}$$

The vector:

$$\dot{\mathbf{A}} = \dot{A}\mathbf{k} = \frac{1}{2}r^2\dot{\theta}\mathbf{k},$$

is called the *areal velocity*.

Next we will derive some useful alternative forms for the basic equations of motion.

Alternative Forms to the Basic Equations of Motion for a Particle in a Central Force Field. Recall the basic equations of motion given previously. They will be our starting point:

$$m(\ddot{r} - r\dot{\theta}^2) = f(r), \qquad (11.8)$$

$$m(r\ddot{\theta} + 2\dot{r}\dot{\theta}) = 0. \qquad (11.9)$$

Previously we derived the following *constant of the motion*:

$$r^2\dot{\theta} = h = \text{constant}. \qquad (11.10)$$

In the problems at the end of this chapter you will show that this constant of the motion will allow you to determine the θ component of motion, provided you know the r component of motion. However, (11.8) and (11.9) are coupled (nonlinear) equations for the r and θ components of the motion. How could you solve them without solving for both the r and θ components? This is where alternative forms of the equations of motion are useful.

Let us rewrite (11.8) in the following form (by dividing through by the mass m):

$$\ddot{r} - r\dot{\theta}^2 = \frac{f(r)}{m}. \qquad (11.11)$$

Now, using (11.10), (11.11) can be written entirely in terms of r:

$$\ddot{r} - \frac{h^2}{r^3} = \frac{f(r)}{m}. \qquad (11.12)$$

We can use (11.12) to solve for $r(t)$, and the use (11.10) to solve for $\theta(t)$.

Equation (11.12) is a *nonlinear* differential equation. There is a useful change of variables, which for certain important central forces, turns the equation into a *linear* differential equation with constant coefficients, and these can always be solved analytically. Here we describe this coordinate transformation.

Let

$$r = \frac{1}{u}.$$

This is part of the coordinate transformation. We will also use θ as a new "time" variable. Coordinate transformation are effected by the chain rule, since this allows us to express derivatives of "old" coordinates in terms of the "new" coordinates. We have:

$$\dot{r} = \frac{dr}{dt} = \frac{dr}{d\theta}\frac{d\theta}{dt} = \frac{h}{r^2}\frac{dr}{d\theta} = \frac{h}{r^2}\frac{dr}{du}\frac{du}{d\theta} = -h\frac{du}{d\theta}, \tag{11.13}$$

and

$$\ddot{r} = \frac{d\dot{r}}{dt} = \frac{d}{dt}\left(-h\frac{du}{d\theta}\right) = \frac{d}{d\theta}\left(-h\frac{du}{d\theta}\right)\frac{d\theta}{dt} = -h^2u^2\frac{d^2u}{d\theta^2}, \tag{11.14}$$

where, in both expressions, we have used the relation $r^2\dot{\theta} = h$ at strategic points.

Now

$$r\dot{\theta}^2 = r\frac{h^2}{r^4} = h^2u^3. \tag{11.15}$$

Substituting this relation, along with (11.14) into (11.8), gives:

$$m\left(-h^2u^2\frac{d^2u}{d\theta^2} - h^2u^3\right) = f\left(\frac{1}{u}\right),$$

or

$$\frac{d^2u}{d\theta^2} + u = -\frac{f\left(\frac{1}{u}\right)}{mh^2u^2}. \tag{11.16}$$

Now if $f(r) = \frac{K}{r^2}$, where K is some constant, (11.16) becomes a linear, constant coefficient equation.

Central Force Fields are Conservative. Now we will show that central forces are conservative forces. We already know that there are many important implications that will follow from this fact, such as conservation of total energy.

If a central force is conservative then the work done by the force in moving a particle between two points is independent of the path taken between the two points, i.e. it only depends on the endpoints of the path. In this case we must have:

$$\mathbf{F} \cdot d\mathbf{r} = -dV$$

where V is a scalar valued function (the potential). Evaluating the left-hand side of this expression gives:

$$\mathbf{F} \cdot d\mathbf{r} = f(r)\frac{\mathbf{r}}{r} \cdot d\mathbf{r} = f(r)dr,$$

where we have used the relation $\mathbf{r} \cdot d\mathbf{r} = rdr$. You can derive this relation by noting that $\mathbf{r} \cdot \mathbf{r} = r^2$, and then computing the differential of this equality. Therefore,

$$-dV = f(r)dr,$$

from which it follows that:

$$V = -\int f(r)dr. \tag{11.17}$$

Hence, if we know the central force field, (11.17) tells us how to compute the potential.

Conservation of Energy for a Particle in a Central Force Field. Since central forces are conservative forces, we know that total energy must be conserved. Now we derive expressions for the total energy of a particle of mass m in a central force field. We will do this in two ways.

First Method. First we compute the kinetic energy. The velocity is given by:

$$\mathbf{v} = \dot{r}\mathbf{r}_1 + r\dot{\theta}\boldsymbol{\theta}_1,$$

and therefore:

$$\mathbf{v} \cdot \mathbf{v} = v^2 = \dot{r}^2 + r^2\dot{\theta}^2.$$

The kinetic energy is given by:

$$\frac{1}{2}mv^2 + V = E.$$

Therefore, we have:

$$\frac{1}{2}m\left(\dot{r}^2 + r^2\dot{\theta}^2\right) - \int f(r)dr = E. \tag{11.18}$$

Second Method. The second method deals directly with the equations of motion and realizes the expression for the total energy as an integral of the equations of motion.

We multiply (11.4) by $\dot{r}$, multiply (11.5) by $r\dot{\theta}$, and add the resulting two equations to obtain:

$$m(\dot{r}\ddot{r} + r^2\dot{\theta}\ddot{\theta} + r\dot{r}\dot{\theta}^2) = f(r)\dot{r} = \dot{r}\frac{d}{dr}\int f(r)dr$$

$$= \frac{dr}{dt}\frac{d}{dr}\int f(r)dr = \frac{d}{dt}\int f(r)dr,$$

or,

$$\frac{1}{2}m\frac{d}{dt}\left(\dot{r}^2 + r^2\dot{\theta}^2\right) = \frac{d}{dt}\int f(r)dr.$$

Integrating both sides of this equation with respect to time gives:

$$\frac{1}{2}m\left(\dot{r}^2 + r^2\dot{\theta}^2\right) - \int f(r)dr = E = \text{constant.}$$

Problem Set 11

1. Prove that in cartesian coordinates the magnitude of the areal velocity is $\frac{1}{2}(x\dot{y} - y\dot{x})$.

2. Derive the equation:

$$\frac{d^2r}{d\theta^2} - \frac{2}{r}\left(\frac{dr}{d\theta}\right)^2 - r = \frac{r^4 f(r)}{mh^2}.$$

3. Show that the position of a particle as a function of time moving in a central force field can be determined from the equations:

$$t = \int \frac{1}{\sqrt{G(r)}}dr, \qquad t = \frac{1}{h}\int r^2 d\theta, \qquad (11.19)$$

where

$$G(r) = \frac{2E}{m} + \frac{2}{m}\int f(r)dr - \frac{2h^2}{m^2 r^2}.$$

4. (a) Find the potential energy for a particle which moves in the force field:

$$\mathbf{F} = -\frac{K}{r^2}\mathbf{r}_1,$$

where K is some positive constant.

(b) How much work is done by the force field in moving a particle from a point on the circle of radius $r = a > 0$ to a point on the circle of radius $r = b > 0$? Does the work depend on the path?

5. Consider the relation:

$$r^2\dot\theta = h = \text{constant},$$

that we derived earlier. Explain how it enables us to determine the θ component of motion if we know the r component of motion.

6. How is the expression $r^2\dot\theta$ related to the angular momentum of the particle about O?

Bibliography

Ernst Mach. *The Science of Mechanics: A Critical and Historical Account of its Development.* Open Court Publishing Company, 1907.

Isaac Newton. *The Principia: Mathematical Principles of Natural Philosophy.* Univ of California Press, 1999.

Arnold Sommerfeld, MO Stern, and RB Lindsay. *Mechanics.* Lectures on Theoretical Physics, Volume i, 1954.

Murray R Spiegel. *Theory and Problems of Theoretical Mechanics*, Schaum's Outline Series. *and Ex*, 3:73–74, 1967.

Clifford Truesdell. *Essays in the History of Mechanics.* Springer Science & Business Media, 2012.

Index

acceleration, 24
acceleration due to gravity, 67
angular momentum, 114
arclength, 24, 28
areal velocity, 123

Cartesian, 13
central force, 119
central force field, 119
central force fields are conservative,
  124
coefficient of friction, 69
components, 15
components normal to the motion, 68
conservation law, 91
conservation of energy for a particle
  in a central force field, 125
conservation of momentum, 94
conservative, 85
conservative force fields, 87
conserved, 91
constrained motion, 68
constructing the phase portrait, 103
coordinate free, 13
coordinate system, xvi, 13
coordinates, 13
cross, 8
cross product in coordinates, 18
curl, 39
curvature, 30

curve, 27
curve in space, 23, 55

derivative of a vector, 20
determinant, 19
difference of two vectors, 4
differentiation of vectors, 20
dimensionless form, 112
dot, 7, 17
dynamics, xvi

energy, 81, 92
energy method, 109
equations of motion for a particle in a
  central force field, 120
equilibrium point, 105, 107
expressing a vector in coordinates, 15

flat Earth, 68
force, xviii, 1, 50
frame of reference, 50
friction, 68

geometrical picture, 102
gradient, 39
gravitational acceleration, 67

inertial coordinate systems, 52

kinematics, xvi
kinetic energy, 85

law of areas, 121
law of conservation of energy, 92
length, 15
level curves, 102
level set of the energy function, 107
level sets, 102
librations, 113
line integral, 35, 45
linear, 57

mass, xvii, xx, 50
mechanics, xv
momentum, xviii, 1
muliplication of vectors, 6
multiplying vectors by scalars, 4

Newton's equation for the pendulum,
    109
Newton's first law, xvii, xviii
Newton's Law of Gravitation, 68
Newton's Laws of Motion, 48
Newton's second axiom of motion,
    55
Newton's second law, xix
Newton's third law, xix
nonlinear, 57

ODE, 57
orthogonal, 7
orthonormal, 7

partial derivatives, 38
partial differentiation, 38
pendulum, 109
periodic solutions of Newton's
    equations, 113
perpendicular, 7
phase plane, 107
phase plane analysis, 101
phase portrait, 107, 111
potential, 88
potential energy, 85, 88
power, 81, 84

projectiles, 67
Pythagorean, 16

reaction force, 68
rectangular, 13
rescaling time, 111
right hand, 8
right-handed coordinate system, 14
rotations, 113

saddle point, 105, 107
scalar, 1, 7, 17
separatrix, 107
space, xvi, xx
stable equilibrium point, 107

3-Tuple, 17
tangent, 28
the classical principle of relativity, 51
the difference of two vectors, 4
the negative of a vector, 4
the product of a vector and a scalar, 5
the projection of a vector onto
    another vector, 7
the sum of two vectors, 4
time, xvi, xx
torque, 113
total energy, 92
turning points, 105, 107

unit binormal, 30
unit tangent vector, 28
unit vector, 5
units, xvi

vector, xviii, 1, 8, 18
vector field, 35, 107
vector valued function, 35, 41
velocity, 23

work, 81

zero vector, 4

Printed in the United States
by Baker & Taylor Publisher Services